U0336533

衷藏雅尚·海上流晖——王水衷捐赠服饰精选

Luminizing and Elegance: Qipao in Shanghai—Selection of Costumes Donated by Wang Shuizhong

上海市历史博物馆 | 上海革命历史博物馆
SHANGHAI HISTORY MUSEUM | SHANGHAI REVOLUTION MUSEUM

上海书画出版社

编辑委员会

Editorial Board

目录

Contents

前言

台北中华文物学会理事长、中国台湾收藏家王水衷先生将其珍藏的338件海派旗袍等服饰与相关饰品，无偿捐赠给上海市历史博物馆（上海革命历史博物馆），表达了他对伊始于沪上的海派旗袍重归故里之美好情怀。为此，上海市历史博物馆精选其中的数十件佳品，举办“衷藏雅尚·海上流晖——王水衷捐赠服饰展”，在感谢捐赠者的同时，旨在回眸海派旗袍对上海近代服装文化所奉献出的时尚与经典，进而诠释上海旗袍海纳百川、追求卓越的核心价值观。

兴起于20世纪二三十年代的海派旗袍，既兼收并蓄了满汉服饰的实用要素，又融会演绎了中西服饰的唯美理念。曾经是“东方巴黎”的上海，无论是名媛、贵人，还是女生、女工，都市女性或多或少的以穿着属于自己的旗袍来表现追求摩登、崇尚自由和婉约雅致的心智情趣。此次展览，上海市历史博物馆力图以学术研究为支撑，以文物、影视和数字技术等多视角向观众讲述海派旗袍背后的上海服饰文化、上海“云裳”品牌、上海匠心制造和上海百货购物的故事。

最后，感谢王水衷先生和所有为此次展览的举办做出不懈努力的人们，并祝展览取得圆满成功。

胡 江

上海市历史博物馆（上海革命历史博物馆）馆长

2018年8月

Preface

Mr. Wang Shuizhong, chairman of the Taipei Chinese Culture and Fine Arts Association and a collector from Taiwan, has donated 338 pieces of Shanghai-style qipao (also known as cheongsam) and related accessories to the Shanghai History Museum (Shanghai Revolution Museum) for free, expressing his kind wishes to return the qipao to their home in Shanghai– the place where they originated. To this end, the Shanghai History Museum holds the exhibition *Luminizing and Elegance: Qipao in Shanghai – Exhibition of Costumes Donated by Wang Shuizhong*, an apparel exhibition donated by Wang Shuizhong, by carefully selecting dozens of the best objects from this collection. While showing gratitude to the donor, the exhibition also aims to recall the fashion and classics Shanghai-style qipao contributed to Shanghai's contemporary costume culture, and in turn, interpret the core values of Shanghai's qipao, namely its great inclusiveness and pursuit of excellence.

The Shanghai-style qipao, starting to flourish in the 1920s and 1930s, not only absorbed and combined the practical elements of Manchu and Han costumes, but also jointly reflected the aesthetic concepts of Chinese and Western costumes. In modern Shanghai, the once " Paris of the East ", urban women, regardless of socialites, celebrities, or schoolgirls, female workers, more or less wore their own qipao as a way of expressing their graceful taste and their pursuing of modern life and freedom. In this exhibition, with the support of academic research, the Shanghai History Museum strives to tell the audience about the Shanghai costume culture behind Shanghai-style qipao, as well as the stories of the Shanghai Yungzong brand, the Shanghai craftsmanship and the Shanghai Department Stores, from a multi-angled view through cultural relics, films and digital technology.

Finally, I would like to express my gratitude to Mr. Wang Shuizhong and those who have made relentless efforts to support this exhibition, and I wish the exhibition a great success.

Hu Jiang
Director of Shanghai History Museum (Shanghai Revolution Museum)
August, 2018

永安有限公司
WINGON CO. LTD

旗袍作为华人女性的时尚服装
被誉为中国国粹和女性国服
由发源地上海风靡至全中国

As a fashionable dress for Chinese women
qipao is known as the quintessence of China and the national costume of women
from Shanghai to all over China

旗丽曼妙

中华民族的伟大复兴受到国际瞩目，特别在日趋坚实的文化软实力上，彰显出中国文化对全球与日俱增的吸引力、号召力及影响力。

上海是全国经济中心，人文荟萃，拥有最具蓬勃活力的中外文化气息，是当今世界上最具影响的国际化大都市之一，也是国家历史文化名城。新落成的上海市历史博物馆位于繁华的市中心，由八十余年的历史建筑改建而成，见证此地风云际会，成为上海市重要文化地标。馆内收藏有大量反映上海发展和历史沿革的珍贵文物和文献，体现出上海城市精神的璀璨辉煌，更具体呈现国家加强文物保护利用、丰富群众性文化活动、弘扬中华优秀传统文化的重要宣示，富有指标性意义。

旗袍作为华人女性的时尚服装，被誉为中国国粹和女性国服，由发源地上海风靡至全中国。旗袍以其流动的旋律、曼妙的风情，表现出中华女性典雅大气、宁静婉约、温柔贤淑的气质，是近现代文明新装，展现出民族健康美，体现出文化自信力，并蕴藏着女性坚毅安定的力量。服饰是时代的重要象征之物。而旗袍文化又是海派文化重要的组成部分，浪漫绰约、经典时尚。旗袍追随着时代，承载着文明，凝聚着情感，尽管潮流随着岁月更迭，但是旗袍光华历久弥新，如今更引起西方时尚服饰的东方热，持续展现文化的风采和永久魅力。

海峡两岸一家亲，同根同文，民族情感长系我心，吾人长年耕耘推动两岸文化交流。悉闻上海市历史博物馆新馆落成，并陆续推出精彩展览，其高度重视历史文化之用心，令人感动敬佩。逢此因缘，身为中华儿女寻思能为此略尽些绵薄之力，承蒙中共上海市委宣传部、中共上海市委台湾工作办公室和上海市文物局的费心安排，本人将自己近三十年来所收藏的20世纪30年代左右之旗袍服饰三百多件，悉数无偿捐赠给上海市历史博物馆收藏，这些旗袍款式优雅多姿，且面料纹样花色齐备，织造工艺细致华美，望之可喜，展现历史风华的韵致。真诚的心意，让文物适得其所，并且在专业的研究团队下进行建档存藏、学术研究、公众展览，再现光彩，相信是这些美丽旗袍的最好归宿。

上海旗袍之美百花齐放，文化自信光辉灿烂。是展为上海市历史博物馆、各位领导及同仁付出辛劳与努力的具体成果，本人有幸得以躬逢盛事，为民族文化尽一份心意备感荣耀，在此谨向各级领导、各界人士、主办单位等致上最诚挚的敬意！也感谢我的朋友和我的家人在我从事公益的时候给予大力的支持，祝福展览圆满成功。

王水衷

台北中华文物学会理事长

The Charm of Qipao

The great rejuvenation of China has attracted international attention. The increasingly solid cultural soft power demonstrates the increasing attraction, appeal and influence of Chinese culture on the world.

Shanghai is the national economic center, the most vigorous cultural city and one of the most influential cosmopolitan in the contemporary world as well as a historical city. Situated in the prosperous downtown area, the newly-built Shanghai History Museum was renovated from a historical building which is more than 80 years old. The building has witnessed the history and became an important cultural landmark. The museum has a large collection of precious cultural relics and documents reflecting the historical transformation of Shanghai's development and revolution, embodies the prosperous spirit of the city, and is the application of the important announcement by the government of strengthening the protection and utilization of cultural relics, as well as enriching mass cultural activities and promoting the excellent traditional Chinese culture.

As a fashionable dress for Chinese women, qipao is known as the quintessence of China and the national costume of women, from its origin, Shanghai, to all over the world. The curvy silhouette and charming design show the elegant and graceful qualities of Chinese women. Qipao presents the beauty of health, confidence of culture and soft but strong mind of women. Costumes are a signature of times, and the culture of qipao is a vital part of Shanghai style culture. Qipao follows the time, carried, the civilization and condenses the emotion. Although the trend may change, qipao remains classic, The Asian fever in western design continues to show the permanent charm of culture.

Taiwan and mainland are one family; we share the same culture and same root. I have been working hard for a long time to promote the cultural exchange between Taiwan and mainland. As I heard the new building of Shanghai History Museum has been built and new exhibitions have been displayed, I admire their intention to value history and culture, and was trying to make some contribution. Arranged by Publicity Department of the CPC Shanghai Committee, the Office of Taiwan Affairs of Shanghai Municipal People's Government and Shanghai Cultural Relics Bureau, I have donated more than 300 pieces of qipao from 1930s to Shanghai History Museum for free. These qipao styles are graceful and varied, with beautiful patterns and meticulous weaving techniques, showing the audience the charm of history. I sincerely believe that it is the best home for those beautiful pieces to be used to carry out archiving and storage, academic research and public exhibition under the professional research team.

The beauty of Shanghai qipao varies, and the light of culture confidence is glowing. The smooth undertaking of this exhibition has been made possible due to the hard work of many leaders and colleagues. I am honored to be part of this grand event and would express my utmost gratitude to all parties involved. I would also like to thank my friends and family for their solid support for my public service. Wish the exhibition every success!

Wang Shuizhong
Director of Chinese Culture and Fine Arts Association

旗袍的发展与海派文化

旗袍的源，也许可以追溯到中国历代的长袍，加上作为限定词的“旗”，说明至少在字面上和清朝的旗人脱不了干系。但是清代文献中并未出现“旗袍”这个称谓。我们费了很大的力量，只在1925年出版的《雪宧绣谱》中找到它。这是一本由刺绣大师沈寿口授，清末状元张謇整理成文的著作。在这本书里，“旗袍”的所指，确实是清代旗人之袍，并不是近代以来中国人心目中的旗袍。

清代旗人不论男女都把袍当作礼服或公开场合中合乎礼仪的服装。旗人自己并不把它叫作旗袍。正如宋人并不把当时流行的几何框架中填以花鸟纹饰的锦称为宋锦。旗袍本来是他者使用的能指。清代汉人男性（包括其他少数民族）也穿袍，与满人并无二致。只有汉女可以穿自己传统的两截装裙襦，即衫、袄、褂等与裙的组合。所以，也可以认为旗袍最早主要是汉人用来指称旗女之袍。

辛亥革命发生，推翻了清帝国，以满人为核心的旗人害怕有人仍然记得几百年前扬州和嘉定发生的屠城事件，害怕这迟来的报复，就纷纷清除身上代表满族身份的象征符号，如改姓易服。所以在1911年至1920年这差不多十年间，男人长袍上的马蹄袖不见了，旗女也不再穿袍，改穿露出裤装的长短褂子。

事实上旗人担忧的大规模民族复仇并没有发生。到了1920年代，穿袍的女性又慢慢多了起来，包括本来不穿袍的汉族女性。大部分地区人们缘于不太远的历史记忆，就把它称为旗袍，而广东等南方城市则称之为长衫。所以至今英语中还可以用长衫的音译作为旗袍的对译。

最初的旗袍就是极大简化了的旗女之袍。除了一个地方，一个人来人往的大城市，呼吸着国际时尚的气息，踩着国际流行的步点，并以此赋予了旗袍新的生命——那就是上海。

在上海，20世纪20年代旗袍有了倒大袖，加入了迪考艺术的很多特征。上海旗袍与建筑，家具、日用器物和书面装帧等等一起，甚至构成了迪考艺术的一个分支，叫上海阿迪克。

从此以后，旗袍的立领高或低，袖的或长或短或无，下摆的位置，开衩的高低，都与国际流行节奏有了高度的相关性。20世纪三四十年代中，旗袍引进了开省道（从腰省到胸省）、装袖笼（替代接袖）、使用揿钮和拉链（替代纽襻）等西式服装的技术。其面料从厚重渐趋轻薄，更多印花，减少织锦和绣花。夏天使用的镂空蕾丝和秋冬使用的针织拉绒（衬绒），这些都是学习了源于西方的纺织品，而且大量使用了合成染料，如阴丹士林蓝。旗袍不仅是民国服制规定的妇女礼服，不仅是大家闺秀贵妇名媛的衣橱必备之品，也是青年学生、纺织女工、家庭妇女和女佣们的日常服装，是中国特色的一件制套裙。旗袍开衩处闪现的不再是长裤、套裤或膝裤，而是长筒丝袜，上海人称之为“玻璃丝袜”，旗袍与各种西式服饰搭配都有浑然天成的效果，如羊毛开衫、绒线开衫（上海女性中有非常多的编结绒线衫的高手乃至大师）、裘皮大衣、海虎绒大衣、手笼和坤包。所以，即使号称专攻西式女装的上海鸿翔，做旗袍照样一流。可以说，以上所有这些变化，都是在上海完成的。旗袍面料、款式、图案和色彩上的变化和创新，都与上海脱不开关系。旗袍不管如何千变万化，都能使人一眼认定这就是旗袍。上海鸿翔女装的传人金泰钧曾经说过，世界上诸如和服、韩服、纱丽、纱笼等民族服装都是以不变应万变的，一变就失去了它们的身份，所以它们永远是经典的传统，不是时装。只有旗袍一直在变，兼具了时尚和传统两种看似无法调和的身份。

新中国成立之初，广大女性追求朝气蓬勃的革命新形象，旗袍一度受到冷落。20世纪50年代中期，在政府的号召下，旗袍以简洁素雅的风格重新受到重视，特别在从事文娱和外交等工作的妇女中流行。改革开放以来，很多设计师都曾致力于旗袍的推广和复兴，但一度并不成功。直至最近的十年中，旗袍才再度流行，在中国高级定制领域更是一枝独秀。其设计和制作的中心仍在上海。曾经为电影明星胡蝶设计旗袍的百岁老人褚宏生，去年逝世前，仍然在上海为旗袍故事续写美丽篇章。

旗袍与艺术领域的海派文化也密不可分。海派画家从张大千、吴湖帆到陈逸飞，都曾创作过不少身着旗袍的女性形象。上海各种报刊的摄影和插画中，也频频出现旗袍女子的身影。上海创作和上演的电影、话剧乃至沪剧中，旗袍女性也常常是不可或缺的亮点。上海作家特别是女作家的笔下，女性的旗袍形象被描绘得栩栩如生，张爱玲有《更衣记》，王安忆有《长恨歌》，程乃珊则有《上海女人》和《蓝屋》等等。

祖国欣欣向荣日益强大，国人的自信心和自豪感也随之增强。旗袍正在成为当代中国女性身份的视觉符号。旗女之袍虽然是古代由北方少数民族挟武功而进入，却是在上海中西融合的文化中得以蜕变而获得新生，并可期望它传之久远。

包铭新

东华大学 教授

2018年6月

The origin of qipao could be traced back to the long gown throughout Chinese history. Although the Chinese character Qi refers to Bannerman or Manchus from the Qing Dynasty, the term qipao cannot be found in Qing literature. We spent a lot of effort and find it only in the 1925 *Xueyi Embroidery* Book. This is a book dictated by Shen Shou, a master of embroidery, and written up by Zhang Qian, a scholar at the end of Qing Dynasty. In this book, the word qipao does refer to 'Manchu robe', rather than what Chinese commonly think of in modern times.

In Qing Dynasty, both men and women wore robes as ceremonial dress or formal dress on public occasions. By that time, the long robes were not called "Manchu robe" by themselves, just as Song people wouldn't call their popular brocade, flower and bird ornament in geometric frame, "Song brocade". Han men (including other ethnic minorities) in Qing Dynasty also wore robes, which were the same as Manchus. Only Han women could wear two pieces for dressing, which were the combination of short top, jacket with skirt. Thus, qipao can be regarded as what Han people used to call Manchu women's robe.

The 1911 Revolution overthrew the empire of the Qing Dynasty, Manchus then eliminated the Manchu identity to avoid being revenged by people who still remembered the massacre in Yangzhou and Jiading hundreds of years ago. Therefore, between 1911 and 1920, the horse-hoof-shaped cuffs on men's long robe were gone, while the Manchu women changed their robes into jacket with pants.

In fact, the massive ethnic revenge never happened. Around the 1920s, long robes enjoyed popularity again, and even attracted Han women who did not wear robes before. Thus most people during that time named it qipao, while some southern regions such as Guangdong called it Changshan. From Changshan derived the loanword cheongsam, which has still been used to refer to qipao in English.

The earliest qipao were just simplified Manchu robe. Apart from one place, where people's aesthetic sense began to change as a result of the influence of Western fashions, and thus giving qipao a new life. This place is called Shanghai.

In Shanghai, qipao introduced wide cuff with bell-shaped sleeves, and added characters of Art Deco in 1920s. Shanghai qipao along with architecture, furniture, daily utensils and book binding, emerged as a branch of Art Deco, called Shanghai Art Deco.

From that period on, the design of qipao including the height of collar, length of sleeves, side slits and hemline were all highly correlated with the international trend. In the mid 1930s to 1940s, Western tailoring features such as bust darts and waist darts, shoulder seams, and metal zippers were introduced in quick session. The fabric gradually became soft and lightweight; brocade and embroidery were replaced with printed pattern. The hollow lace used in summer and the knitted velvet (lining velvet) used in autumn and winter were also textiles originated from the West. Synthetic dyes such as indanthrene were widely used. qipao was not just 'Women's Formal Wear' stipulated in Clothing Regulations promulgated by the Nationalist government in 1929, not just for ladies from privileged families, but also became daily wear for Young students, textile workers, housewives and maids, as one-piece-dress of Chinese identity. Under the side slits, there were no longer trousers or pants, but fashionable stockings, which Shanghai people called 'glass stocking'. Qipao also matches perfectly with any Western outfit, such as woolen cardigan, knitted cardigan (there were many masters of knitting among Shanghai people), fur coat, fleece coat and handbag. Therefore, even Shanghai Hongxiang, which specialized in western style women's clothing, was able to make first class qipao. The changes and innovations in qipao fabrics, styles, patterns and colors were inseparable from Shanghai. The nature of qipao never changes, no matter how much change it made. Jin Taijin, descendant of Shanghai Hongxiang once said, other ethnic costumes such as kimono, hanbok, sari and sarong were never majorly modified, so they are always regarded as classic traditions, rather than fashion. Only qipao has been changing, with two seemingly unreconciled identities of fashion and tradition

At the beginning of the founding of new China, the majority of women pursued the vigorous new revolutionary image, and qipao was once neglected. In the middle of the 1950s, under the call of the government, qipao was re-valued because of its simple but elegant style. It regained popularity among women engaged in the work of entertainment or diplomacy. Since the reform and opening up, many designers have devoted themselves to the promotion and revival of qipao, but it has not been successful for a while. Until the recent ten years, qipao seems be revived and stands out in Chinese Haute Couture field, mainly based in Shanghai. Zhu Hongsheng, a centenarian who designed qipao for movie star Hu Die, had been writing beautiful chapter for qipao his whole life until the last minute before he passed away last year.

Qipao related closely also to the Shanghai style culture in the art field. Shanghai style painters such as Zhang Daqian, Wu Hufan and Chen Yifei have all created female images in qipao. Women in qipao also appeared frequently on various newspaper and magazines. The image of women in qipao is depicted vividly in Shanghai writes' writing, especially female writers such as Chinese Life and Fashions wrote by Zhang Ailing, *Everlasting Regret* wrote by Wang Anyi, Shanghai Woman and Blue House by Cheng Naishan etc.

As China is growing stronger and more prosperous, the confidence and pride of Chinese people are getting stronger. qipao is becoming a visual symbol of the identity of contemporary Chinese women. The long robe was brought by Manchus in the ancient time but metamorphosed in Shanghai's fusion of Chinese and western culture, and is expected to pass on forever.

Bao Mingxin
Professor of Donghua University
June, 2018

纺织面料与旗袍创新

所谓旗袍，经常被人们说成是旗人之袍。虽然旗袍的诞生或与旗人之袍有关，但旗袍的形成和发展主要是在民国初年的上海，从此之后，旗袍的样式，特别是在服装结构和制作工艺上未有大变，而其变化主要集中在纺织面料的设计上，包括纤维材料、技术工艺、组织品种和纹样图案上。纺织面料的变化，对历史上旗袍的发展起到了关键作用，在当下的旗袍作为非物质文化遗产进行传承和创新，也有着极为重要的关系。

一、纤维原料的变化

我国传统的纺织原料是天然纤维，棉、毛、麻、丝。对于女性高档服饰来说，多采用丝绸，特别是旗人之袍，多用绫、罗、绸、缎再加刺绣印染工艺。但有时也有普通的长衫或长袍，只是用棉布制成。到民国时期，旗袍的原料构成开始发生极大的变化。

棉布是当时世界上最为流行的纺织面料。虽然中国也生产土布，但旗袍多采用进口的机织棉布。机织棉布品种很多，但做旗袍的棉布多以所用染料或色彩命名，如阴丹士林布、海昌布、克力登布、爱国蓝布等。属于斜纹的棉布料有棉哔叽、线呢等。

用羊毛材料纺织而成的毛呢也是当时的新型面料。上海开埠后，西洋的呢绒也传入中国，包括粗纺呢绒和精纺呢绒。20世纪20年代起，上海开始开设采用进口毛纱织造精纺呢绒的工厂，生产如哔叽、华达呢、花呢、派立司、直贡呢、凡立丁、啥维呢等。从保存下来的民国旗袍看，一些高档旗袍也采用了薄型呢绒，加上印花处理，呈现出不同于传统的洋派风格。

而在旗袍最为重要的丝绸面料上，材料的变化来自人造丝的应用。1925年，杭州纬成公司开始悄然试织人造丝绸——巴黎缎，大受市场欢迎。由于人造丝的成本远低于厂丝，因而获利颇丰，引起了同行的效仿。到1928年杭州的人造丝用量已上升到57%，超过了桑蚕丝的用量。随着纬成公司巴黎缎的成功，各地亦开始生产人造丝织物，其中苏州的巴黎缎、花襄绸和盛泽的中山葛名噪一时；1937年上半年，在上海丝绸产品中，全人造丝织物已达21.92%，人造丝交织物多至70.74%，而真丝织物仅占6.48%。

二、技术工艺的变化

旗人之袍的面料全部采用手工工艺，大多是在绫罗绸缎等提花织物上的刺绣，少量的彩绘和印染。而民国时期技术和工艺的变化，则主要来自织造技术的提高和印染工艺的进步。织造技术的提高首先是新型织机的发明。1800年前后，法国丝织业重镇里昂的贾卡发明了新式的提花织机，用带纹板的提花龙头替代用线综编制花本的束综提花机。此后，贾卡提花机向欧洲各地及亚洲、特别是日本传播。到民国初年，浙江杭州又从日本引进了贾卡织机，当时称为手拉机，并逐渐在丝织行业中推广使用。此后的20世纪二三十年代，江浙沪等地进而又引进电力丝织机，建立了近代丝织工厂，使得丝织生产率大大提高。

正是因为在原料上并用厂丝和人造丝，设备上采用高效率的电力织机，使得一些如高捻度、起绉、起绒、袋组织、呢地、多梭箱换道等现代织造技术可被用于当时的丝绸面料生产，诞生了一大批在传统木织机上不可能生产的新品种。在春夏季有大量的薄型绉类、纱类产品，在秋冬季有大量锦类、缎类、葛类产品。机器生产的织锦缎，因为采用了人造丝、提花龙头和双梭箱装置，花色灿烂而成本下降，成为大众穿得起的奢侈品。

此外，丝绸印染技术也得到了快速的发展。首先是化学合成染料的进口和生产，为丝绸带来无穷的新的色彩，使得旗袍面料更为多彩多韵。同时，机器印花的出现使得新设计的纹样图案能以较快的速度、较低廉的价格、更为精美的品质生产出来，所以，20世纪二三十年代前后，印花丝绸一时大量涌现，成为当时新的时尚。与此同时，机器练染工厂也在上海等地出现，丝绸精练逐步改用平幅精练，而精练技术的提高促进了大量生织匹练品种的生产，如双绉、素软缎等，也提升了印花织物的品质。

三、组织品种的丰富

清代常见的丝绸品种有各种绫罗绸缎，或是织锦妆花之类。到民国时期，旗袍的面料依然以绸缎居多，但同样是绸缎，但已和清代的面料有了很大的区别。传统面料品种单一，代表性的有摹本缎（以南京所产最多）、库缎（通过正反缎纹显花）、宁绸（斜纹地上起斜纹花）、花线春（平纹地上起斜纹花）、花纱（有亮地纱、实地纱与芝地纱之分）、杭罗（平纹与绞纱配合起横、直条纹）、杭纺（平纹素绸，同类型的有盛纺、绍纺等，以产地命名）、湖绉（以湖州所产最多）等。

而旗袍面料同样是丝绸，品种却极为丰富，有直接从国外进口的，进口面料与时尚花色精彩纷呈，在审美理念上越出了传统，向西方时尚靠拢。有仿进口面料的，如被称为泰西缎或泰西纱的，这类品种的纹样循环很大，特别是纹样的过渡区的组织变化极为丰富，明显是用贾卡织机提花才能生产的。此外还有仿其他纤维的，甚至几种纤维交织的，层出不穷。如利用真丝和人造丝染色性能差异进行生织套染的双色或三色新品种如克利缎、鸳鸯绉等，全人造丝的提花织物明华葛，其光亮赛如蚕丝绸匹。此外，还利用酸性溶液对人造丝和桑蚕丝溶解作用的不同而生产烂花绒。

这里特别要指出的是蕾丝或是针织物。今天所见的民国旗袍中，经常见到针织旗袍，其中最女性化、最性感的是蕾丝旗袍。蕾丝也称花边织物，19世纪末引进中国，是手工编织的缕空织物，而用于旗袍的蕾丝是经编机织物。鉴于当时中国尚未掌握其生产工艺，蕾丝面料应该是进口的。上海女性在当时穿着蕾丝旗袍，即使衬上里布，这份勇气也着实可嘉。在所有的旗袍面料中，蕾丝传递给我们的是一种别样的海上风情。

四、纹样花色的时尚

旗袍面料最为引人入胜的是其纹样图案，现代丝绸设计中称为花色。这些花色让人眼睛一亮，我们看到的是一种时代新风、现代气象。旗袍面料的花色体现在以下几方面：

首先是大量的花卉纹样，其中有着少量对称的花卉纹样，但更多是自然写实主义的意境。它们可以是清地大朵花，也可以是满地小碎花，可以是传统的梅兰竹菊，也可以是玫瑰与郁金香。还时常会看到小几何纹作地，带着些装饰艺术的味道。还有不少花卉经过了变形处理，不追求花卉的形似，从而为纹样题材的丰富打开了大门。

旗袍面料的另一主题是条格、圆点等几何、抽象纹样，并配以各种色彩的变化。这一时期与美国的机械工业设计风格基本相吻，这与机器时代追求简洁明快的风尚有关。作为现代设计的要素之一，没有比醒目的几何纹样便能传递出直观的秩序之美了。除了规矩的几何纹，还可以将其变形，或采用肌理纹样，或看似随机线条的乱涂，使旗袍花色充满时代的美感。

第三是外来纹样的流行。比如佩兹利纹样，被上海设计师形象地称为“火腿纹样”。再如20世纪二三十年代流行的大型花卉纹样、孔雀羽纹样，其卷曲的线条有着欧洲新艺术风格的特征，而20世纪40年代带几何装饰感的流行纹样，又让人联想到法国起源的装饰艺术风格。

上海是近代中国的经济中心，集工业、商业、消费为一体，也成为时尚的中心。到时下的21世纪，旗袍又一次成为人们关注的具有中国特色的女性服饰，但日常生活中的旗袍并不多见，而特殊场合穿着的旗袍风格又十分单一，或落俗套，或在表演时的显得十分重工。如何形成丰富的旗袍文化，是人们正在探索的目标，而纺织面料的创新则是其中一个关键的因素。无论是纤维原料还是技术工艺，无论是组织品种还是纹样花色，都需要在传承的基础上加快创新。现在，大上海提出的是科技旗袍，由东华大学这一所纺织学科一马当先的高等学府举旗呐喊，而杭州市举办的是邀请全球名家设计旗袍，希望产生时尚的、独特的、能融入生活的旗袍，其关键点恐怕都在于纺织面料的创新。上海历史博物馆能在此时收藏和展出王水衷先生捐赠的旗袍精品，让人们观摩近百年来旗袍的发展中纺织面料的作用，非常有益。

（此文中有较多部分参考袁宣萍和徐铮的相关文章，特别致谢！）

赵　丰

中国丝绸博物馆馆长

2018年7月

Qipao was commonly known as Manchu robe. Although the origin of qipao might be traced back to Manchu robe, the formation and development of qipao were mainly in Shanghai during the early years of the Republic of China. Since then, there has been no significant change in the design of qipao, especially in the silhouette and tailoring. The progresses have been made in the areas of fiber material, production technology, weaving and pattern. The change of textile fabrics has played a vital role in the development of qipao in history. It also has a strong connection with the inheriting and innovating of qipao, which is an intangible cultural heritage now.

1. The change of raw material of fiber

Our traditional textile materials are natural fibers, cotton, wool, linen and silk. For luxurious clothing for women, silk is most commonly used. Especially Manchu robes, the majority of which were made by different types of silk as twill damask, Luo gauze and satin with the decoration of embroidery. But sometimes there are ordinary long shirts or gowns, just made of cotton as well.

By the period of the Republic of China, the material composition of qipao began to change greatly. Cotton was the most popular material for clothing. Although China also produces hand woven cloth, qipao normally uses imported cotton cloth woven by machine. There are many kinds of machine-made cotton cloth, but the cotton used for making qipao is usually named after the dyes or colors used in Chinese, for instance indanthrene cloth, Hydron cloth etc. Twill cotton fabrics during that time were cotton serge, twine cloth etc.

Woolen cloth woven from wool was also the new fabric of the time. After Shanghai opened its port, western wool was also introduced into China, including woolen wool and worsted wool. From the 1920s, Shanghai opened a factory that used imported woolen yarn to weave worsted wool, producing fabric like serge, gabardine, palace, venetian, valetine and semifinish serge. From the preserved qipao of the Republic of China, some high-grade qipao also adopted tropical suiting, with printing pattern, presenting a different Western style.

On the case of silk, the most important fabric to make qipao, the change of materials comes from the application of artificial silk. In 1925, Hangzhou Weicheng Company began to quietly try out the artificial silk——Paris satin, which was quite popular in the market. Because the cost of rayon was much lower than that of the factory reeled silk which leads to the rise in profit, causing the imitation of the peers. By 1928, the amount of rayon used in Hangzhou had risen to 57 percent, exceeding the amount used by silkworms. In the first half of 1937, the total rayon fabric in Shanghai reached 21.92 percent, and the number of rayon interlaces reached 70.74 percent, while silk fabric only accounted for 6.48 percent.

2. The change of technology

The Manchu robes were all hand-made, mostly embroidered on silk and other jacquard fabrics, with a small amount of colored drawing and dyeing. The change of technology in the Republic of China mainly came from the improvement of weaving technology and the progress of printing and dyeing technology. The first improvement in weaving technology was the invention of a new type of loom. Around 1800, Jaka of Lyons, a French silk industry town, invented a new jacquard loom, replacing the bunched jacquard loom with a jacquard faucet with a thread harness. Since then, the jacquard machine has been spreading to all parts of Europe and Asia, especially Japan. In the early years of the Republic of China, Hangzhou from Zhejiang Province introduced the jacquard loom from Japan, which was then called the hand-drawing machine and was gradually popularized in the silk weaving industry. In the 1920s and 1930s, electric silk looms were introduced in Jiangsu, Zhejiang Province and Shanghai, and modern silk weaving factories were established, which greatly improved the productivity of silk weaving.

It was because of the use of factory reeling silk and rayon, and the application of high efficiency power looms, some modern weaving techniques such as high twist, crepe, fleece, circular weave, cloth floor and multi-shuttle box channeling could be used in silk fabric production at that time, and various new fabrics which were impossible to make using traditional wooden looms. In spring and summer there were a large number of thin crepe, yarn products, and in autumn and winter a large number of brocade, satin, poplin products. Machine-made brocade, with its colorful design and low cost, has became an affordable luxury for the masses thanks to the use of rayon, jacquard taps and double-shuttle boxes.

In addition, silk printing and dyeing technology had also been rapidly developed. The first was the import and production of chemical synthetic dyestuff, which brought numerous new colors to silk and made qipao fabric more colorful. Meanwhile, the appearance of machine printing made the new design able to be produced at a faster speed, lower price, and more exquisite quality. Thus, around 1920 to 1930, the market was filled with printed silk and became a trend. At the same time, boil-dying factory appeared in areas around Shanghai, the improvement of scouring technology promoted the production of a large number of raw materials, such as crepe de chine and plain satin, also enhancing the quality of printed fabrics.

3. Various weaving type

The typical silks of the Qing dynasty were twill damask, Luo gauze, satin, and brocade satin. In the period of the Republic of China, the fabric of qipao was still mainly silk and satin, but it was already very different from that of the Qing dynasty. Traditional fabric has limited kinds, with the representative of Mopen (popular around Nanjing area), jacquard satin, nankin, Hangzhou silk gauze, and Hangzhou silk plain etc.

Qipao fabric is also silk, but with large variety. Some are imported directly from abroad, which have gone beyond the tradition in the aesthetic concept and close to western fashion; some are imitated imported fabrics, such as cotton venetians, the pattern circulation of which is rather long, especially the structure change of the transitional zone of pattern is extremely rich, easy to tell was produced by jacquard loom. In addition, there are fibers imitating other fibers, even several fibers interwoven, and numerous combinations.

For instance, two colored or tricolor crepe satin; all rayon jacquard fabrics Minghua Poplin, are brighter than silk. In addition, the different dissolution effects of acid solution on rayon and silkworm silk were used to produce burnout fabric.

It was worth mentioning on lace or knitwear. In Republic of China qipao passed to today, we often see knitted qipao, among which the most feminine and sexy one is lace qipao. Lace, was introduced to China at the end of the 19th century as a fabric woven by hand, while the lace used in qipao was warp knitting fabric. Since China had not yet mastered its production process, lace fabric should be imported. Shanghai women at the time wore lace qipao, even with lining cloth inside, the avant-garde spirit is really commendable. In all the qipao fabrics, the lace passed to us is a different kind of fusion.

4. Trend of pattern

The most attractive point of qipao textile is its pattern, which draws attention at first glance. The design and color of qipao fabric are reflected in the following aspects.

First of all, there are a large number of patterns of flowers, which have a small number of symmetrical patterns of flowers, but more naturalistic realistic artistic conception. They could be big flower on plain background, or allover floral. They could be traditional plum orchid bamboo chrysanthemums, or roses and tulips. Also there are small geometric pattern. There are many flowers that have undergone deformation treatment and do not follow the shape of flowers, which is mind-blowing.

Another theme of qipao fabric is geometric and abstract patterns such as stripes and dots, and with a variety of color changes. This period is the same time as the machinery industry in the United States, which is related to the pursuit of simplicity fashion in the machine age. As one of the elements of modern design, no other patterns can convey the intuitive beauty of order better than geometric patterns. In addition to the regular geometric patterns, it can also deform, or use texture patterns or random lines to make qipao full of the aesthetic feeling of the times.

The third group is imported pattern. For example, paisley pattern which Shanghai designers knew as "Ham pattern". Another example is the large flower pattern and peacock feather pattern popular in the 1920s and 1930s, whose curly lines have the characteristics of European new art style. In the 1940s, the popular pattern with geometric decoration also reminds people of the decorative art style (Art Deco) originated from France.

Shanghai is the economic center of modern China, which integrates industry, commerce and consumption, and also becomes the center of fashion. In the 21st century, qipao has once again become a focus of attention for women's clothing with Chinese characteristics. However, qipao is rarely seen in daily life, whilst the style of qipao on special occasion is very simple or tacky or overdressed. How to form qipao culture is the target that the mass are exploring, and the innovation of textile fabrics is one of the key factors. No matter material or technology, weaving or pattern, they all require speeding up innovation on the basis of inheritance. Nowaday, what Shanghai proposes is the science and technology qipao, by Donghua University, a leading institution in the field of textiles, whilst Hangzhou is inviting famous designers from all over the world to design qipao in the hope of producing fashionable, unique and life integrated qipao. Perhaps the key point still lies in the innovation of textile fabrics. The exhibition hold by Shanghai History Museum made it possible for people to observe the vital part textile has played in the hundred years progress of qipao, and is very helpful.

(Acknowledge: With special thanks to Yuan Xuanping and Xu Zheng as many parts of this article referred to the related articles of them.)

Zhao Feng
Director of China National Silk Museum
July, 2018

从女性视角解读海派文化

那大概是七月里的一天，张爱玲穿着丝质碎花旗袍，色泽淡雅，也就是当时上海小姐普通的装束。

——柯灵《遥寄张爱玲》

在上海市历史博物馆筹建的过程中，我们收到了中国台湾收藏家王水衷先生慷慨捐赠的三百余件海派旗袍。这批旗袍不属于特殊的群体或个人，其中也没有"弹眼落睛"之作——事实上，"弹眼落睛"并不是能让上海人心有戚戚的风格。她们属于1930-1940年代万千上海小姐的生活日常，面料、纹样和款式大都精细考究，充满了时尚气息。比如，选用先提花后印花的面料来制作，缘饰、盘扣与旗袍的颜色、纹样遥相呼应，有心者自能体会到穿着者的优雅、精致、低调。又如，时髦的西式纹样和缘饰显示着与20世纪初的欧洲同步的流行节奏，这样的设计也是海派旗袍独有的风味。

女性的生活方式和社会地位，映射着社会的文明程度和文化性格。旗袍为我们提供了一个绝佳的女性视角，从这个视角我们看到1930-1940年代的上海社会：已显露出现代城市文明的开放和个性，也保持着传统中国的风骨和韵致。因此，我们策划了一个旗袍风尚展，用文化的线索来贯穿整个展览，希望观众踏入展厅与穿着这些精致旗袍的上海小姐一同，经历一趟摩登上海的文化之旅。

展览由序幕和主体构成。序幕选择旗女之袍、旗袍马甲、西式连衣裙这三个与海派旗袍密切相关的服饰典型，为最热闹的海派旗袍风尚拉开帷幕。上海是中国经历现代国家转型的主要阵地之一，传统和现代的元素，本土和外来的信息，渗透在20世纪初的服饰变化之中。序幕旨在营造这样的背景氛围，务求使观众一目了然，心中有底。

展览的主体分为三幕。第一幕"时尚盛宴 文化沙龙"，通过不同时尚舞台上旗袍风尚的视觉盛宴，展示其背后的文化力量。

开篇是时装公司和百货公司里的旗袍风尚，这是可以直接观看、感受和参与的商业文化。上海史专家熊月之教授认为海派文化具有趋利性或商业性、世俗性或大众性、灵活性或多变性、开放性或世界性四大特点，其中最根本的是趋利性。（熊月之《"海派文化"的得名、污名与正名》）商业或趋利是一种有节制的导向，使旗袍的设计更贴合人的生活习惯和审美趣味。著名画家叶浅予设计了一款格纹短袖旗袍，长度至小腿中部，开叉及膝。他写道："目下最流行的是长旗袍，可是你在马路上走时，或下车上车时，你一定觉得她太长了吧？而且你再仔细想想晚上穿着的情形，你就觉得她非常文雅幽娴并能衬出女性的娇态了……所以这里拟定半长式的旗袍，专为下午或出门之用。此式垂到小腿中段为正确之长度，出手约一尺，袖宽三寸半，最大是四寸，领可低些，如能采用斜方格图案衣料，则更合于时令。”（叶浅予《最流行之新装》，1931年第18期《玲珑》）既有出于穿着者习惯的考虑，也以美术家独特的时尚触觉，引导大众的审美。

有学者认为，民国时期的时装是美术家时代的时装，相较与巴黎、纽约，这一时代服装最具中国特色的核心即是美术。（周松芳《民国衣裳：旧制度与新时尚》）画坛名流如叶浅予、方雪鸪、张乐平、江小鹣等都扮演过服装设计师的角色，在报刊上刊登设计款式。我们选择了一些当时的时尚杂志所刊登的设计图，同时也选择了一些与设计图相似的旗袍和具有印象派画风的旗袍实物，来做相应的展示。这一美术引导时尚的特色也体现在时装表演上。上海是中国最早举办时装表演的城市，早期的时装表演常常出现在游艺会、慈善演出中，带有寓美术教育于表演的目的。（《联青社筹办大规模游艺会》，1926年11月22日《申报》）自1930年代以后，鸿翔、朋街、永安、先施等服装或百货公司竞相举办专场时装表演，推介当季时装，中国的时装表演才算是走上与西方国家相似的以商业为目的的道路。

展览以云裳服装公司的旗袍故事为例，对上述文化现象做出集中展示。云裳服装公司于1927年8月7日开幕，是一家由沪上文人、社交名流和美术家合作开设的服装公司。开幕的当日，《上海画报》《北洋画报》都有大篇报道，稍后包天笑又作了《到云裳去》一文，详细介绍公司情况，刊发于《晶报》上，是一时盛事。公司的股东包括徐志摩、周瘦鹃、胡适、张禹九等文化名流。知名女性张幼仪、唐瑛、陆小曼与公司的经营都有关系。（陈建华《陆小曼·1927·上海》）云裳早期的美术设计师是留法雕塑家江小鹣，后期为著名画家叶浅予。1927年10月云裳参加汽车展览会，表演新装展示，其中不少新装即是旗袍。1930年10月9日上海市第三节国货运动大会上有两场旗袍表演，所展示的旗袍均来自云裳。借旗袍这一视角，我们看到云裳是一次时装实业的创业，更是一场上海文化的沙龙。

1930–1940年代的上海不仅是时尚的中心，也是传媒的中心。旗袍时装是各类报刊的热门话题，大型综合画报《良友》《永安月刊》都刊登过梳理旗袍发展演变的专题，而女性刊物《玲珑》则经常刊登旗袍设计漫画。值得一提的是当时出现了许多妇女杂志，以良友旗下的《妇人画报》和《玲珑》为代表，不仅关注女性的妆容和服饰，经常发布欧洲的流行资讯，也关注女性的生活方式和进步思想，探讨“什么是真正的摩登”的话题。

第二幕“现代女性　摩登生活”，以风情各异的旗袍，讲述不同类型的摩登女性关注自我，追求美好生活的故事。

将旗袍与女性典范相结合是近年来较为流行的展陈方式。无论何种历史文物，相较于社会大众所服用的物品，名人的物品更容易获得保存和珍视。它们可能具有卓越的艺术价值和特殊的历史价值，这样的展览自然精彩。但正是由于其精英属性，使她们很难成为社会大众的代表和缩影。我们将批量的旗袍常服放回到20世纪三四十年代，研究穿着旗袍的人群，比对记录旗袍的文字和影像，发现喜爱穿着旗袍的女性是有一些共性的，她们在不同程度上直接或间接受到了新文化新思想的影响，关注自我的需求——有对美丽形象的需求，有对精致生活的需求，有对精神追求的需求。同时，她们身上仍有一些传统的精神，有些成为成就她们的文化根性，有些却成为埋葬她们的一抔黄土，后者的存在使转型时代显得更为完整，也无须避讳。

因此，我们选择了一些穿着旗袍的新女性的代表，她们有的出身于名门望族，有的出身于中产家庭，有的出身于社会底层。她们或生长于上海，或在斯处短暂地居住、逗留。她们穿着海派旗袍的形象依然闪耀在银幕里，渲染在画布上，刻画在书卷中。

第三幕“民族工业　上海制造”，通过旗袍制作技法的发展和旗袍面料、纹样的变化，讲述中国传统的与西方现代的文化因子在上海交织为海派经典的故事。

旗袍制作是充满人文关怀的传统手工技艺，她体现穿着者的需求和审美。在传统社会，女性的审美主要由男性主导。一方面是美要符合礼制的规定，有经典的依据。《牡丹亭·闺塾》中有一幕，杜丽娘欲为师母绣鞋上寿，向老师陈最良请个鞋样。陈最良说：“依《孟子》上样儿，做个‘不知足而为屦’罢了。”不仅审美趣味，即身体特征也尽量依照礼制去塑造、改变，或走个象征性的程序。当然，其中也包含了对身体特殊部位的讳言。另一方面，美直接来源于男性本身的好恶。《玉台新咏》和《花间集》是集中收录描摹女性体态诗词的代表，所收诗词美则美矣，然大多是以男性视角书写闺怨、宫怨，审美趣味则以绮靡浮艳为主。至于“楚王好细腰，宫中多饿殍”所指的极端病态的审美趣味亦俯拾皆是，不必赘言。

对女性身体特征的尊重和基于这种尊重而产生的健康向上的审美观，是20世纪三四十年代上海所具备的现代文明特征之一。1928年7月15日第76期《常识》报刊文提到“今日上海之女性，已十九解放矣，往来于道者，大都双峰高耸……男女双方均已司空见惯，实无羞耻之理。”这类宣传女性解放胸部是为健康、卫生和美感的文字常见于报刊。而在当时的各类广告、照片中，经常出现穿着西式服装露出健美体态的形象。女性的身体特征受到尊重，自然的体态获得欣赏，这是受西式审美影响而产生的，是西方文化在上海的本土化。这是当时的审美趋向。

旗袍表现女性的曲线之美正是顺应这种审美趋向的结果。1930年代中后期至1940年代，旗袍的制作开始融入了归拔、省道等技法，从适应人体的曲线，到表现曲线的美感。适度的贴合着人体自然曲线来剪裁和缝制，而不是将身体塞进某种标准的曲线尺寸里，是海派旗袍自信的风度。这些文化的变化写在旗袍师傅的技艺之中，正因为并不是他们有意去表现，才更昭示了社会文化的集体无意识。

进入20世纪后，上海的民族纺织工业实际已相当发达。服装面料的织法十分精细，纹样相当丰富，在一定程度上也受到西式工艺的影响。我们选取了具有代表性的两点——Paisley纹样旗袍和阴丹士林蓝布旗袍来展示，前者显示了当时上海的流行时尚实与欧洲同步，后者则是外来工艺与传统因子结合，转化为上海独有的文化符号的代表。

旗袍展是展览策划的热门选题，策展或注重旗袍的发展演变，或讲究旗袍的制作工艺，或表现旗袍的造型美感，或书写名媛旗袍的高贵典范。蔡襄《士伸知己赋》云:“匪衷藏之雅尚，羌得志而弗为。”我们认为海派旗袍之所以成为经典，正由于她的内里是时代的风尚和文化的力量。因此，我们回避了这些已然为观众呈上许多精彩展览的角度，尝试用文化的线索来展示旗袍风尚，有心者若反其道而观之，借旗袍这一女性视角，解读海派文化的精彩，也未尝不可。

历史的长河望不见尽头，文化的海洋浩瀚无边，如果这样的尝试能如石子般激起一些水花，或引来更为精妙的金玉之作以飨观众，便是无负关心此事的各方人士的眷眷之意了。

张　霞

上海市历史博物馆（上海革命历史博物馆）馆员

2018年6月

It was about one day in July, Eileen Chang was wearing a silk Qipao with delicate floral design and elegant color, which was common attire for a Shanghai lady at that time.

—— Ke Ling Miss Eileen Chang

During the process of establishing the new hall, we have received a generous donation of more than 300 pieces of Shanghai Style qipao from a Taiwan Collector. These qipao did not belong to any special groups or individuals, and among which there were no eye-catching pieces. In fact, "eye catching" is not something echoes with Shanghai people. They belong to the daily lives of millions of Shanghai ladies in the 1930s and 1940s. Their fabric, pattern and style are mostly exquisite and fashionable. For instance, the fabric is first jacquard woven then printed; Fringes and frogs are matched with the color and pattern. Those who pay attention would be able to feel the elegant, exquisite and understated style of the owner. As another example, the stylish western pattern and fringe on qipao illustrate that the same fashion trend as Europe also existed in Shanghai in the early 20th century. This shows a unique character of the Shanghai style.

The lifestyle and social status of women reflect the degree of civilization and cultural character of the society. Qipao provides us with an excellent female perspective, from which we can view the Shanghai society in the 1930s and 1940s: not only revealing the openness and individuality of modern urban civilization, but also maintaining the traditional Chinese vigor and charm. Thus, we have organized this qipao fashion exhibition with cultural clues throughout, with the hope that the audience could experience a cultural trip to modern Shanghai together with the Shanghai ladies who wore these exquisite qipao.

The exhibition consists of a prologue and a main part. In the prologue we chose some Manchu robes, long vests and western one-piece dresses as three types of typical Shanghai style costume to begin with. Shanghai is one of the main positions where China has experienced the transition to modern country. The traditional and modern elements, local and foreign information permeated the change of clothing in the early twentieth century. The purpose of the prologue is to provide such background information so that the audience can have a basic knowledge about the whole story.

The main body of the exhibition is divided into three parts. The first chapter "fashion feast, cultural salon" displays the concealed cultural power behind the visual feast of different fashion stages.

The beginning is the qipao fashion in fashion companies and department stores, which is a commercial culture that can be seen, felt and participated in directly. Professor Xiong Yuezhi, an expert in Shanghai history, believes that the culture of the Shanghai style has four characteristics, which are profit-orienting or commercial, secular or popular, flexible or variable, inclusive or global, and the most fundamental one is commercial (*Naming of Shanghai Style* by Xiong Yuezhi). Commercial or profit seeking is a restrained guidance that makes the design of qipao more closely fit people's customs and aesthetic tastes. The famous artist Ye Qianyu used to design a short-sleeved plaid qipao, whose hemline went up to the mid-calf and with split up to the knee. He wrote, "Nowadays long qipao is the most fashionable style, but when walking or getting into a car, you must feel it too long. And if you think about wearing it at night, you'll see that she's very elegant and feminine...Therefore, the half-length qipao is specially designed for the purpose of afternoon or outdoor use. So the hemline to mid-calf is the correct length, and the sleeves should be 1/3 meter long, 12-13.2 cm wide, with a lower neckline, ideally using plaid fabric to suit the season ("*The New Popular Style*" by Ye Qianyu on *Linloon Magazine*, Issue 18th , 1931). The design takes care of practical usage, and also uses the unique sense of an artist to guide the public aesthetic.

Some scholars believe that the fashion in the period of the Republic of China was the fashion of the artists. Compared with Paris and New York, the main Chinese character during that time was art ("Clothing of the Republic of China: Old system and New fashion" by Zhou Songfang). Famous Painters such as Ye Qianyu, Fang Xuehu, Zhang Leping, Jiang Xiaojian had all played the role of fashion designers, and published design sketches in newspapers and magazines. We have selected some of those sketches from that time as well as similar qipao to the design to make corresponding display. The tendency of Art leading fashion could also be noticed in fashion shows. Shanghai was the earliest city in China to have fashion show, and it first appeared in recreation fairs and charity performances, with the purpose of artistic education. After 1930s, Hongxiang, Pengjie, Yongan, Xianshi and many more fashion store or department stores began to hold fashion show as a promotion mean to introduce seasonal collections. Thus, Fashion show in China started to become commercialized just as western countries.

Taking the qipao story of Yongzong clothing company as an example, the exhibition presents the above cultural phenomena in a concentrated way. Yongzong clothing company, which opened on August 7, 1927, is a clothing company jointly established by Shanghai literati, social celebrities and artists. On the opening day, *Pictorial Shanghai*, and *The Peiyang Pictorial News* both gave massive coverage, and later on Bao Xiaotian wrote an in-depth article about the company, published on *The Crystal*, creating a major event during that time. The shareholders of the company include cultural celebrities such as Xu Zhimo, Zhou Shoujuan, Hu Shi and Zhang Yujiu. Famous women Zhang Youyi, Tang Ying and Lu Xiaoman are also related to the operation of the company. The graphic designer of the company was first Jiang Xiaojian, a sculptor who studied in France, and then became the well-known painter Ye Qianyu. In October 1927, Yongzong participated in the automobile exhibition, displaying their new collection, many of which were qipao. From the perspective of qipao, we could witness fashion industry's pioneering work, as well as a salon of Shanghai culture.

Shanghai in the 1930s and 1940s was not only the center of fashion, but also the center of media. Qipao fashion was a hot topic in all kinds of newspapers and magazines. Comprehensive pictorials *The Young Companion* and *Wing-on* had both published a feature story on the development of qipao, while *Linloon Magazine* published qipao design quite often. It is worth mentioning that many woman's magazined appeared at that time not only focused on women's makeup and costume, often released popular information in Europe, but also focused on women's lifestyle and their progress of thinking, to discuss the topic of "what is true modern".

The second chapter "Modern Lady, Modern Life" tells the story of modern women who paid attention to their inner-self and pursued a beautiful life.

To combine qipao with female paragons is a popular way of display in recent years. No matter what the historical object is, those owned by celebrities were more easily preserved and cherished compared with what ordinary people had. They might have outstanding artistic and historical value, which makes the exhibition more attractive. But because of their elite attributes, it is difficult for them to become the representatives and epitome of the public. We traced batches of qipao back to the 1930s and 1940s, studied the crowed who wore qipao, compared the words and recordings back then and discovered some common features of the people who were fond of qipao. They were directly or indirectly affected by the new culture and new thoughts, and paid more attention to the needs of themselves - the demand for the beautiful appearance, the demand for exquisite life, and the demand for spiritual pursuit. At the same time, they still kept some traditional spirit in themselves. Some became their cultural root, but some led to their failure. The existence of the latter makes the transition period seem more complete and need not be avoided.

Thus, we have selected some representatives of modern women who were wearing qipao, and were from different social classes. They either grew up in Shanghai or stayed shortly in Shanghai. The images of them in Shanghai style qipao have been still sparkling on the screen, painted on canvas and depicted in books.

Chapter three, "National industry, Shanghai Production", tells the story of how Chinese traditional and western cultural factors interwove to become the classic Shanghai style through the development of qipao's production techniques and the change of fabric and pattern.

Qipao production is a traditional handicraft full of humanistic care. It is considerate of the needs and aesthetics of the wearer. In traditional society, women's aesthetics were mostly dominated by men. On the one hand, beauty should conform to the rules of the courtesy system and be based on tradition. Not only aesthetic taste, physical features should also be shaped, altered, or carried out even in a symbolic manner in accordance with the rules of etiquette, which included as well the denial of specific parts of the body. On the other hand, the definition of beauty came directly from men's taste. Classical poetry from that time were mainly written from male's perspective and the aesthetic tastes were unrealistic. The pursuing of extreme and morbid taste was everywhere.

The respect for the physical characteristics of women and the healthy and upwards aesthetic view based on this respect , were one of the characteristics of modern civilization in Shanghai in the 1930s-1940s. *Common Sense*, issue 76th which was published on July 15th, 1928 wrote "The majority of Shanghai female today are liberated now, the ones we see on the street are all showing their natural curve. There is no reason to be ashamed as both men and women are so accustomed." This type of text promoting the emancipation of women's breasts for health, hygiene and beauty was commonly seen in newspapers and magazines. And in all types of commercials and photos, the figures were often in western fashion and showing a vigorous image. Women's physical characteristics were respected and their natural bodies were appreciated, which was influenced by western aesthetics and the localization of western culture in Shanghai. This was the trend of that time.

The design of qipao showing the curvy beauty of women was the result of this aesthetic trend. From the middle and late 1930s to the 1940s, the production of qipao began to integrate techniques such as the ironing and stretching technique and dart, from adapting to the curve of human body to expressing the beauty of curve. To tailor qipao according to human body rather than fit the body into some standard shape and size, is the confident demeanor of Shanghai style. These cultural changes were written in the skills of the qipao tailors, and it is precisely because they did not intend to show that they reveal the collective unconscious of social culture.

After entering the twentieth Century, the national textile industry in Shanghai had been quite developed. The weaving method of clothing fabric was very delicate, the pattern was plenty, and to some extent, it was affected by the western technology. We have selected two representative examples to display – Paisley pattern and indanthrene. The former shows that Shanghai's fashion at that time was in line with that of Europe, while the latter was the combination of foreign technology and traditional factors, which transformed into Shanghai's unique cultural symbol.

Qipao is a popular topic for exhibition among curators. Some of the curation pay attention to the development of qipao, some study the production techniques of qipao, some show the beauty of the silhouette, and some explore the elegance of celebrity qipao. We believe that the reason why the Shanghai style qipao became classic is due to the fashion of the age and the power of culture that lie within it. Thus, we avoided these existing angles, trying to use culture as a clue to express qipao fashion. You may also use qipao as a female perspective to read the splendid culture of the Shanghai style.

In the long river of history, we hope our attempt can be a little stone to cause some splash, or to attract more exquisite masterpieces in the future, then it would be all we could wish for and hope it is worth all the concern from everyone.

Zhang Xia
Museologist of Shanghai History Museum (Shanghai Revolution Museum)
June, 2018

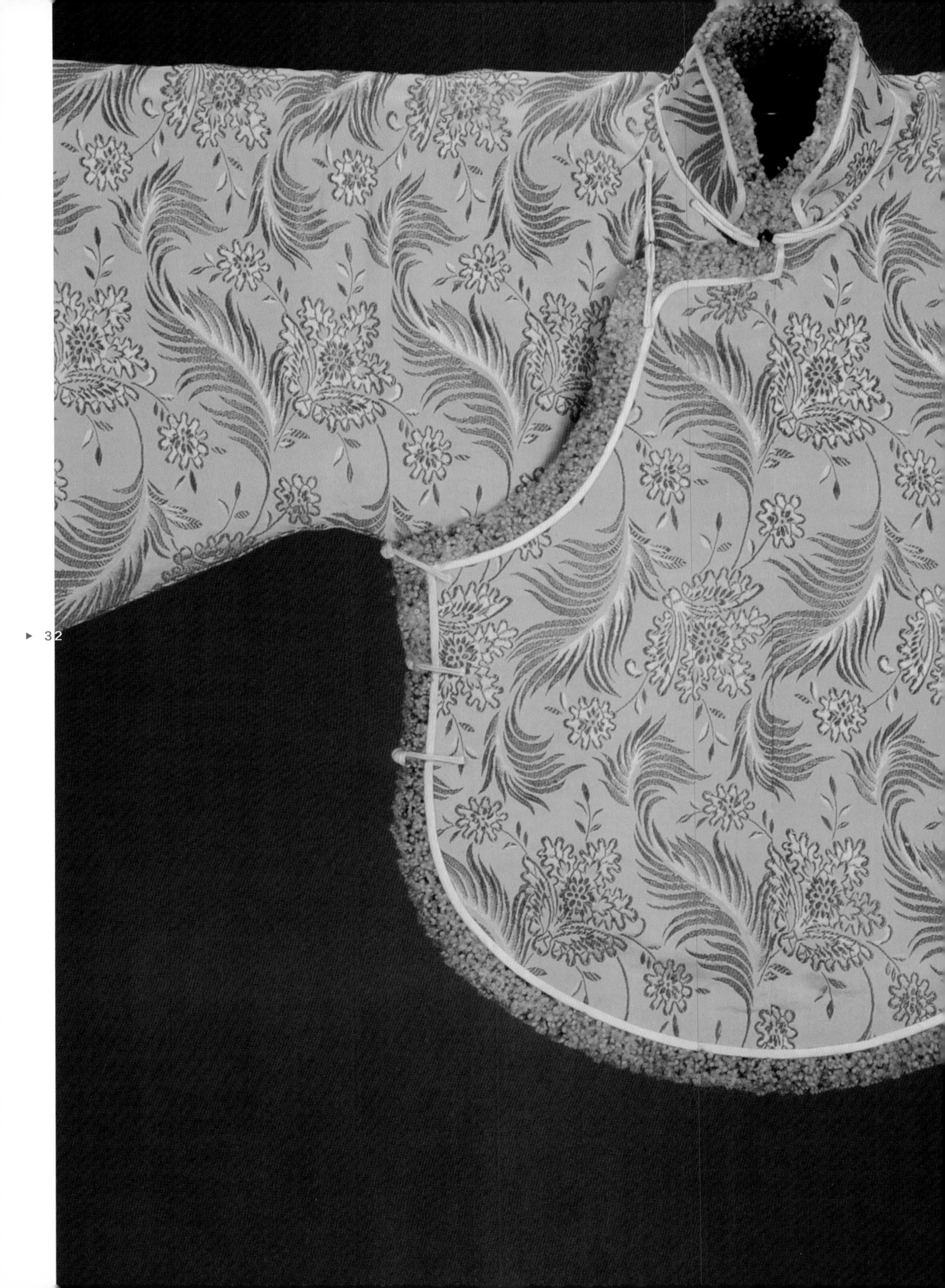

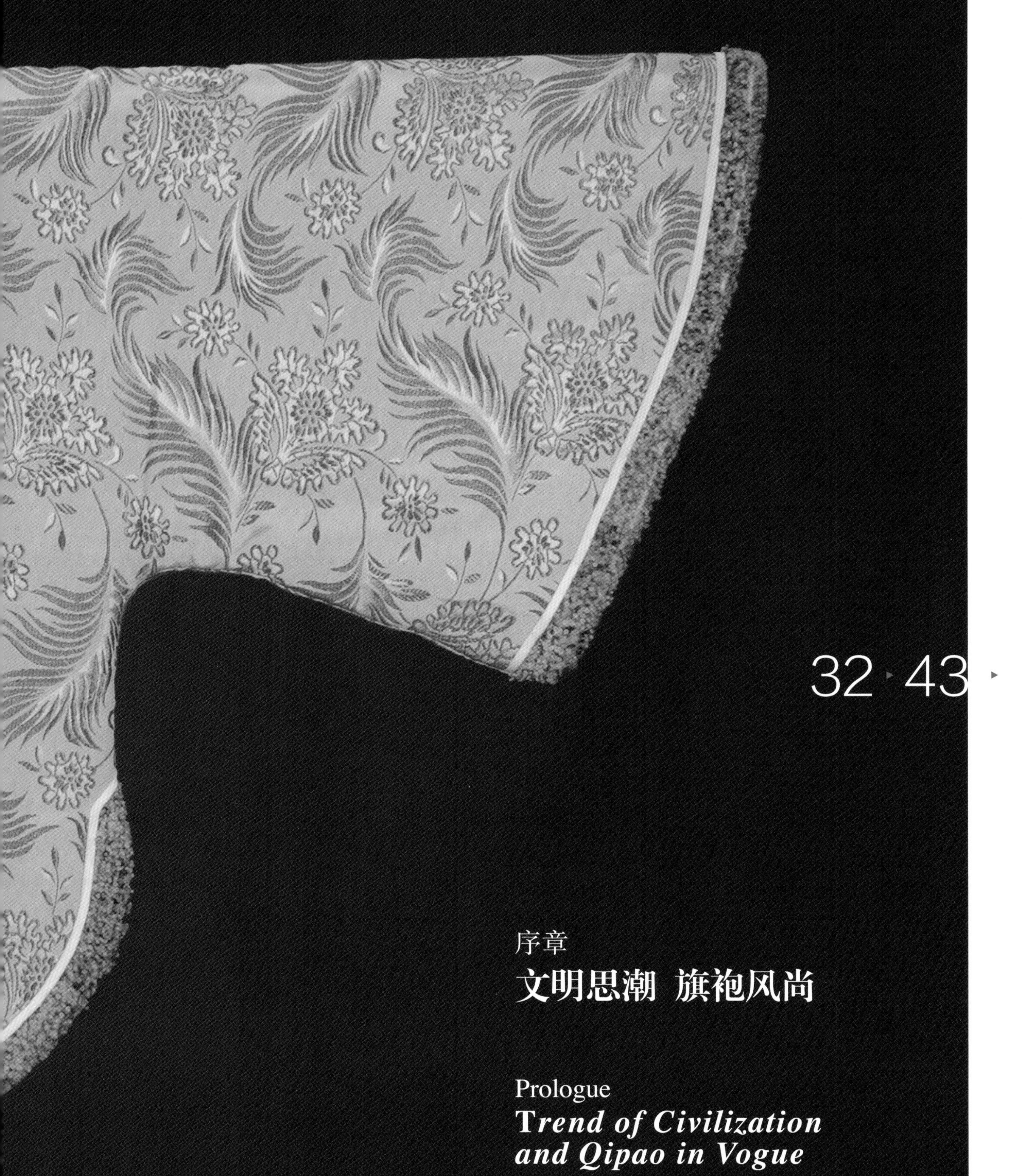

序章

文明思潮 旗袍风尚

Prologue

Trend of Civilization and Qipao in Vogue

序章　文明思潮　旗袍风尚

旗袍风尚始于新文化运动后，是在西风东渐带来的新思潮之下应运而生的时装风尚。在20世纪20年代前后，一些穿着打扮流行元素的交织为旗袍的诞生埋下了伏笔。这些元素有些是不可磨灭的传统印记，有些是伴随着外来思潮而登陆上海的新事物，它们互相融合，成为以服饰为代表的传统文化寻求现代出路的探索。

旗袍的名称和外观与清代旗女袍服有一定关系。旗女之袍大襟右衽，高领宽袖，长及脚面，大都饰以宽阔缘边。袍身左右开衩，穿着时内穿长裤，长裤外有套裤，裤腿绣花，在开衩处若隐若现，对美的表达含而不露。汉族妇女则以上袄下裙的装束为主。

旗袍马甲长及脚背，加在短袄上。穿着这样的马甲，有时底下仍然着裤，有时只穿一双针织袜，配上高跟皮鞋。

西式连衣裙是衣裳连属的一件制，与传统中式袍服的穿用方法完全不同。在上海街头流行的西式连衣裙主要有夏季穿用的日常连衣裙和社交场合穿用的西式礼服。

The fashion of qipao started from the New Culture Movement, and it was a new fashion trend influenced by the Western culture. Around 1920s, the interweaving of some fashionable elements had foreshadowed the birth of qipao. Among these elements, some were traditional style inherited from the past, others were new born characters with external trends of thought. They merged and integrated, becoming the exploration of the modern way out of traditional culture represented by costumes.

The name and look of qipao are related to the Manchu robe to some degree. The latter has a large front lapel opening to the right, stand-up collar with wide sleeves; its hemline reaches the foot, fringed with decorative bands. Manchu robe has slits on both sides, so a pair of trousers had to be worn underneath, with loose and large over trousers where the lower edges have embroidery figures that could be exposed when sitting down or walking. The most common apparel for the Han women were coat with skirt.

The hemline of the long-vest has reached foot, worn over the blouse. Wearing long-vest as such, there should be pants underneath, sometimes a pair of knitted socks and heels.

One-piece dress is a kind of clothing that combines top with bottom. The way of wearing it is completely different from traditional Chinese robe. The fashionable western one-piece dress in Shanghai mainly included summer time daily one-piece dresses and gowns for social occasions.

桃红色团花纱镶花边袍　清代

衣袖展长：122厘米
衣长：136厘米

Peach floral roundel gauze robe with decorative edging *Qing Dynasty*

Length of both sleeves: 122cm
Length: 136cm

米色蕾丝连衣裙 民国（上海市历史博物馆藏）

One-piece dress with beige lace
Republic of China (1912-1949)
(collection of Shanghai History Museum)

黄色提花缎镶花边倒大袖上袄
约1920年代

领：立领，高5厘米
衣袖展长：94厘米
衣长：45厘米
胸宽：38厘米
摆宽：40厘米
扣：盘扣5对，分布在领口1对，
斜襟1对，侧门襟3对

Yellow damask blouse with bell-shaped sleeves
Around 1920s

Collar: Stand-up Collar, 5cm
Length of both sleeves: 94cm
Length: 45cm
Chest width: 38cm
Hemline: 40cm
Button: 5 pairs of frog colsures, 1 on collar, 1 on slant opening, 3 on the side

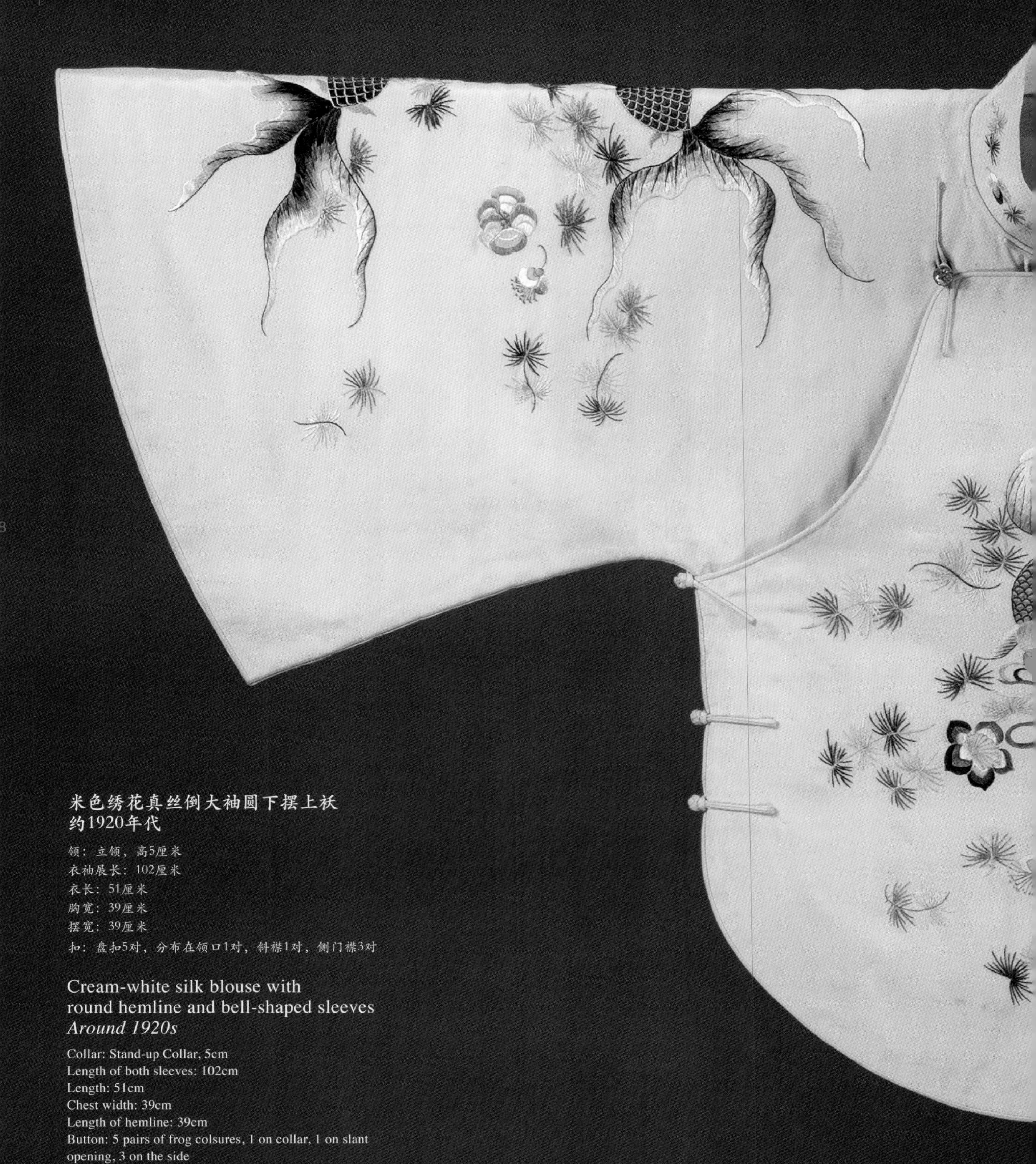

米色绣花真丝倒大袖圆下摆上袄
约1920年代

领：立领，高5厘米
衣袖展长：102厘米
衣长：51厘米
胸宽：39厘米
摆宽：39厘米
扣：盘扣5对，分布在领口1对，斜襟1对，侧门襟3对

Cream-white silk blouse with round hemline and bell-shaped sleeves
Around 1920s

Collar: Stand-up Collar, 5cm
Length of both sleeves: 102cm
Length: 51cm
Chest width: 39cm
Length of hemline: 39cm
Button: 5 pairs of frog colsures, 1 on collar, 1 on slant opening, 3 on the side

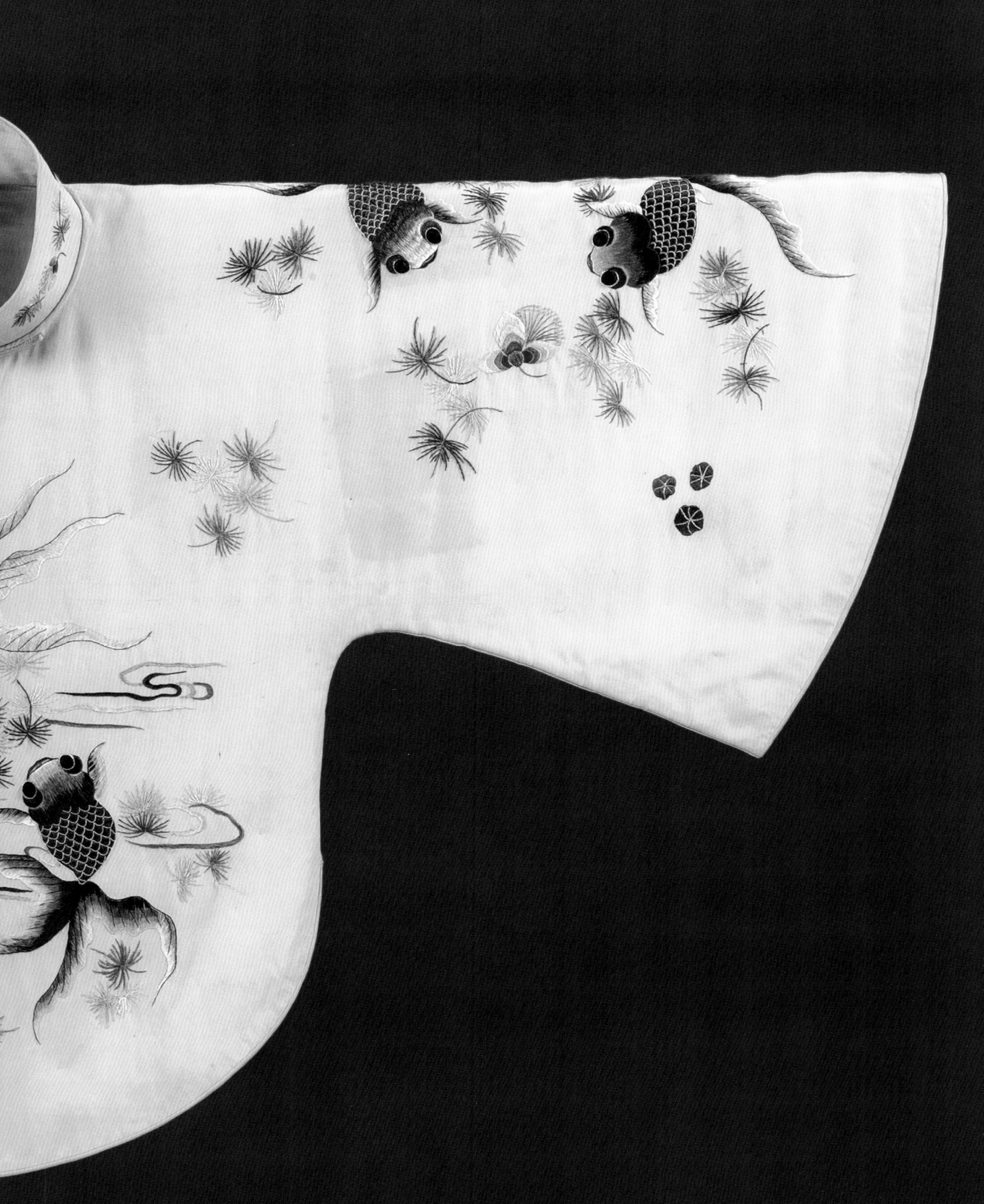

提花印花真丝倒大袖上袄
约1920年代

领：立领，高5厘米
衣袖展长：104厘米
衣长：52厘米
胸宽：37厘米
摆宽：43厘米
扣：盘香扣6对，分布在领口2对，斜襟1对，侧门襟3对

Silk blouse with printed pattern and bell-shaped sleeves
Around 1920s

Collar: Stand-up Collar, 5cm
ength of both sleeves: 104cm
Length: 52cm
Chest width: 37cm
Hemline: 43cm
Button: 6 pairs of frog colsures, 2 on collar, 1 on slant opening, 3 on the side

黑色蕾丝倒大袖圆下摆上袄
约1920年代

领：立领，高5厘米
衣袖展长：101厘米
衣长：52厘米
胸宽：38厘米
摆宽：39厘米
扣：盘扣5对，分布在领口2对，斜襟1对，侧门襟2对

Black lace blouse with bell-shaped sleeves and round hemline
Around 1920s

Collar: Stand-up Collar, 5cm
Length of both sleeves: 101cm
Length: 52cm
Chest width: 38cm
Length of hemline: 39cm
Button: 5 pairs of frog colsures, 2 on collar, 1 on slant opening, 2 on the side

桔色提花真丝倒大袖圆下摆上袄
约1920年代

领：立领，高5厘米
衣袖展长：100厘米
衣长：54厘米
胸宽：40厘米
摆宽：40厘米
摆宽：40厘米
扣：盘扣5对，分布在领口1对，斜襟1对，侧门襟3对

Orange patterned silk blouse with bell-shaped sleeves and round hemline
Around 1920s

Collar: Stand-up Collar, 5cm
Length of both sleeves: 100cm
Length: 54cm
Chest width: 40cm
Length of hemline: 40cm
Button: 5 pairs of frog colsures, 1 on collar,
1 on slant opening, 3 on the side

咖啡色提花镶花边长马甲
1920年代或略早

领：立领，高5厘米
肩宽：29厘米
衣长：117厘米
胸宽：45厘米
腰宽：45厘米
摆宽：65厘米
扣：盘扣6对，分布在领口2对，侧门襟4对；揿纽2对

Coffee patterned long-vest with trimming
1920s or earlier

Collar: Stand-up Collar, 5cm
Shoulder: 29cm
Length: 117cm
Chest width: 45cm
Waist: 45cm
Length of hemline: 65cm
Button: 6 pairs of frog colsures, 2 on collar,
4 on slant opening, 2 metal press studs on the side

咖啡色提花长马甲
1920年代或略早

领：立领，高4厘米
肩宽：25厘米
衣长：110厘米
胸宽：39厘米
腰宽：39厘米
摆宽：60厘米
扣：盘扣4对，分布在侧门襟；揿纽7对

Coffee patterned long-vest
1920s or earlier

Collar: Stand-up Collar, 4cm
Shoulder: 25cm
Length: 110cm
Chest width: 39cm
Waist: 39cm
Length of hemline: 60cm
Button: 4 pairs of frog closures on slant opening,
7 metal press studs

44 ▸ 95 ▸

第一章
时尚盛宴　文化沙龙

Chapter One
Fashion Feast Cultural Salon

第一章　时尚盛宴　文化沙龙

民国时期的上海是远东最大的工业、金融、贸易和文化中心，素有“东方巴黎”的美誉，不仅巴黎当季的时装不数月就出现在上海街头，西方国家的文化和生活时尚也同步融入上海市民的日常。海派旗袍是中西合璧的时装，而西式的生活时尚为旗袍的发展和传播推波助澜。

Shanghai was the largest industrial, financial, trade and cultural center in Asia during the Republic of China, and had the reputation of "Paris of the East". Not only did Paris' seasonal fashion appear on the streets of Shanghai within a few months, but western culture and lifestyle were also integrated into the daily life of Shanghai residents. Shanghai style qipao is a fashion merged by the East and the West, while the western lifestyle added fuel to the development and spreading of qipao.

Chapter One
Fashion Feast, Cultural Salon

一、商店里的摩登服饰

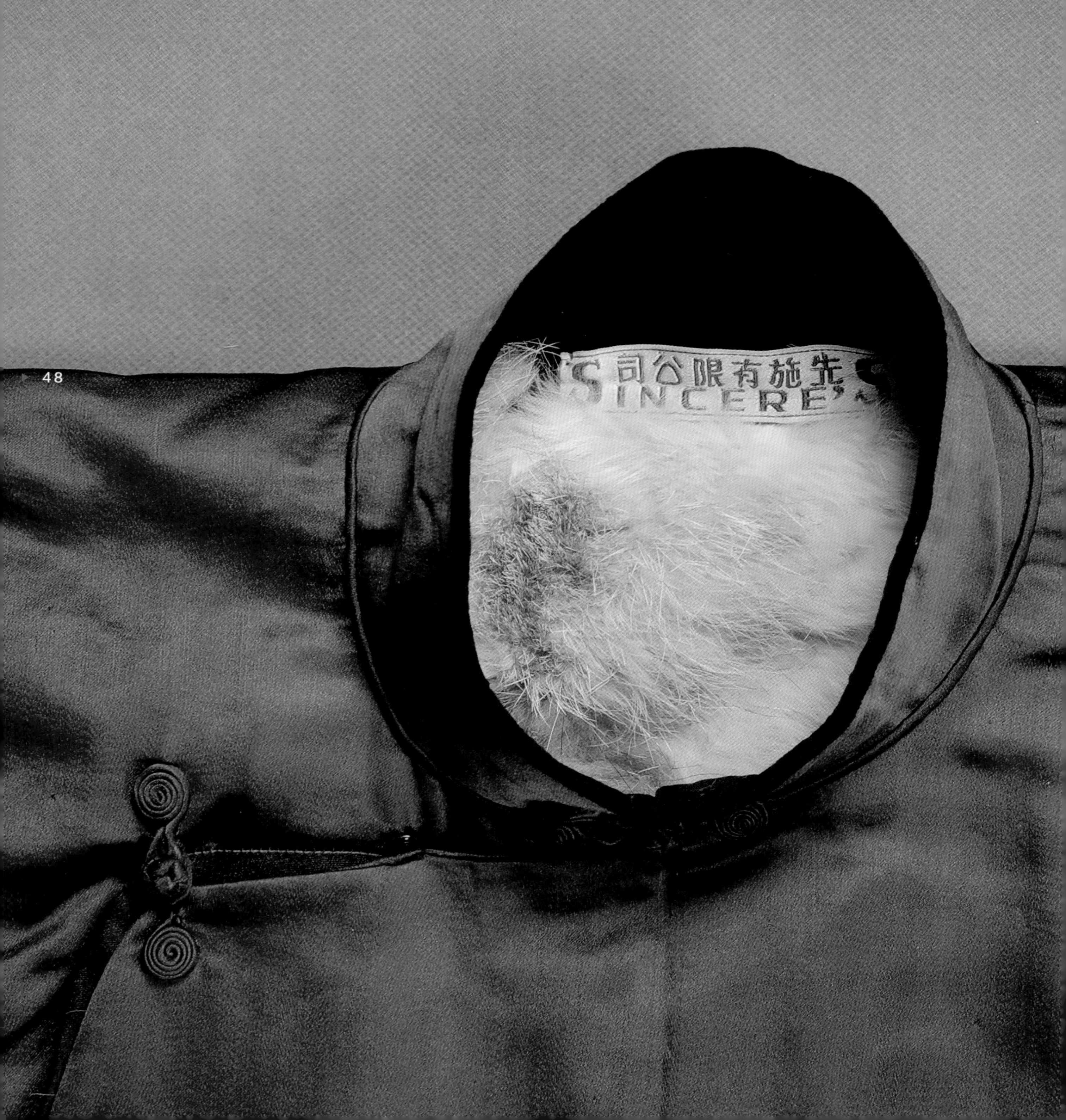

One
Modern Clothing in Department Store

先施公司
The Sincere & Co., Ltd.

一、商店里的摩登服饰

旗袍是百货公司、绸缎公司、女装公司中的热销品类，各大商家竞相推出款式新颖、做工考究的旗袍，可谓各有千秋，各擅胜场。

先施公司于1917年开业，是南京路四大公司中最早开业的一家，也是华人在上海建造的第一家经营环球百货的大型商店。先施公司以"始创不二价，统办环球货"为宗旨，开创了商品明码标价、划一不二的先河。在其楼内还建有旅馆和游乐场，使先施公司成为集购物、住宿、游乐于一体的商业中心，开创了一种新的商业模式。

One: Modern Clothing in Department Store

Qipao is a popular item in department stores, satin companies and women's costume companies. All major sellers enthusiastically launch new designed and delicately made qipao.

The Sincere & Co., Ltd. Opened on 1917, was earliest among the four major companies along Nanjing Road, also the first department store in Shanghai which sold global productions built by Chinese. The policy of the Sincere & Co., Ltd. was "fixed price, global goods", which started trend of non-negotiable price. In the building, there were hotel and amusement park, which made Sincere & Co., Ltd. a business center integrating shopping, accommodation and amusement, and created a new business model.

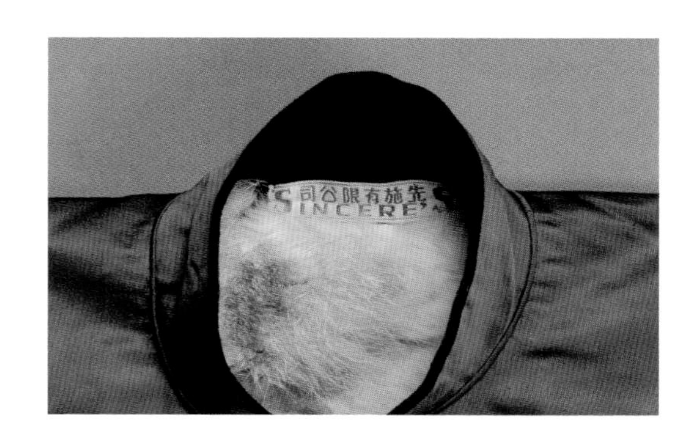

紫色裘里真丝长袖夹旗袍（先施公司）
约1940年代

领：立领，高5.5厘米
衣袖展长：118厘米
衣长：118厘米
胸宽：48厘米
腰宽：47厘米
摆宽：60厘米
侧开衩：22厘米
扣：盘扣9对，分布在领口2对，斜襟1对，
侧门襟6对；揿纽2对

Purple silk long-sleeved qipao with fur lining
(The Sincere & Co., Ltd.)
Around 1940s

Collar: Stand-up Collar, 5.5cm
Length of both sleeves: 118cm
Length: 118cm
Chest width: 48cm
Length of hemline: 60cm
Side slits: 22cm
Button: 9 pairs of frog closures, 1 on collar, 1 on slant opening,
6 on the side; 2 metal press studs

永安公司
The Wing On Co., (Shanghai) Ltd.

永安公司于1918年开业，与先施公司隔街相望，经营环球百货，公司楼内也开设旅馆和游乐场，同先施公司形成竞争之势。

The Wing On Co., (Shanghai) Ltd. opened at 1918 right acrossed street to the Sincere & Co., Ltd., also ran global business and had hotel and amusement park inside, vied with Sincere & Co., Ltd..

红色衬绒提花缎长袖夹旗袍（永安公司）
约1940年代

领：立领，高3厘米
衣袖展长：129厘米
衣长：114厘米
胸宽：44厘米
腰宽：39厘米
摆宽：47厘米
侧开衩：18厘米
扣：盘扣3对，分布在领口1对，斜襟1对，侧门襟1对；揿纽6对

Red damask long-sleeved qipao with velvet lining (The Wing On Co., [Shanghai] Ltd.)
Around 1940s

Collar: Stand-up Collar,3cm
Length of both sleeves: 129cm
Length: 114cm
Chest width: 44cm
Waist: 39cm
Length of hemline: 47cm
Side slits: 18cm
Button: 3 pairs of frog closure, 1 on collar, 1 on slant opening, 1 on the side; 6 metal pressed studs

满地提花缎无袖单旗袍（新新公司）
约1940年代

领：立领，高3厘米
肩宽：39厘米
衣长：121厘米
胸宽：39厘米
腰宽：38厘米
摆宽：46厘米
侧开衩：28厘米
扣：盘花扣2对，分布在领口1对，斜襟1对；盘扣1对，在侧门襟；揿纽6对

Floral damask sleeveless qipao
(The Sun Sun Co., Ltd.)
Around 1940s

Collar: Stand-up Collar, 3cm
Shoulder: 39cm
Length: 121cm
Chest width: 39cm
Waist: 38cm
Length of hemline: 46cm
Side slits: 28cm
Button: 2 pairs of floral frog closuresâ,
1 on collar, 1 on slant opening,
1 pair of frog closures on the side;
6 metal press studs

新新公司
The Sun Sun Co., Ltd.

新新公司于1926年开业，其经营模式与先施、永安相同，规模略小，以开设“玻璃电台”著称。

The Sun Sun Co., Ltd. opened at 1926, the business mode was similar to Sincere & Co., Ltd. and the Wing On Co., (Shanghai) Ltd., but smaller, famous for its “glass radio”.

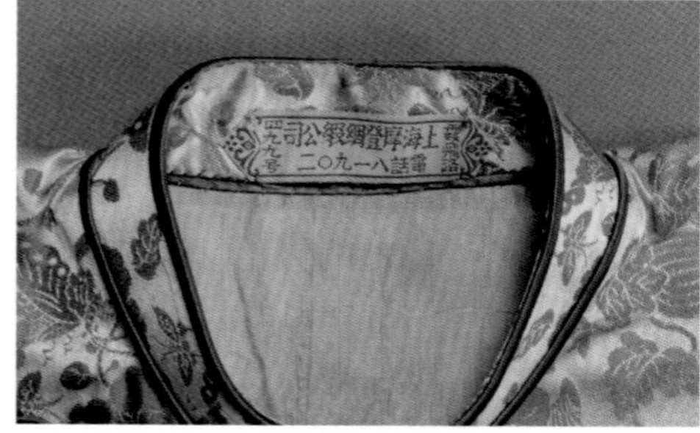

绿色提花缎无袖单旗袍（摩登绸缎公司）约1940年代

领：立领，高2.5厘米
肩宽：43厘米
衣长：112厘米
胸宽：40厘米
腰宽：38厘米
摆宽：43厘米
侧开衩：20厘米
扣：盘花扣2对，分布在领口1对，斜襟1对；
盘扣1对，在侧门襟；揿纽2对

Green damask sleeveless qipao (Shanghai Modern Silk Co.) *Around 1940s*

Collar: Stand-up Collar, 2,5cm
Shoulder: 43cm
Length: 112cm
Chest width: 40cm
Waist: 38cm
Length of hemline: 43cm
Side slits: 20cm
Button: 2 pairs of floral frog closures, 1 on collar, 1 on slant opening, 1 pair of frog closures on the side;
2 metal press studs

蓝色印花真丝长袖夹旗袍（皇后绸缎公司）
约1940年代

领：立领，高4.5厘米
衣袖展长：143厘米
衣长：111厘米
胸宽：45厘米
腰宽：39厘米
摆宽：44厘米
侧开衩：12厘米
扣：揿纽5对

Blue silk long-sleeved qipao with printed pattern and lining (Queen's Silk Co.)
Around 1940s

Collar: Stand-up Collar, 4,5cm
Length of both sleeves: 143cm
Length: 111cm
Chest width: 45cm
Waist: 39cm
Length of hemline: 44cm
Side slits: 12cm
Button: 5 metal press studs

Two
Atelier Yang kweifei

云裳公司开幕式（载《北洋画报》1927年8月27日）
Opening ceremony of Yungzong
(News on The Peiyang Pictorial News, August 27th 1927)

云裳公司开幕式（载《上海画报》1927年8月12日）
Opening ceremony of Yungzong
(News on Pictorial Shanghai, August 12th 1927)

二、云想衣裳花想容

1927年8月7日，在卡德路（今石门二路）静安寺路（今南京西路）路口，一家由沪上文人、社交名流和美术家合作的服装公司开幕，取名"云裳"，吴湖帆以篆书题写公司的中文招牌。"云裳"的典故出自李白描写杨贵妃的著名诗句"云想衣裳花想容"，寓意了公司以艺术改造时装、美化女性的愿景。

Two: "Atelier Yang Kweifei"

On August 7th, 1927, at the crossing of Carter Road (Shimen No.2 Road today) and Bubbling Well Road (West Nanjing Road today), a clothing company Yungzong jointly established by Shanghai literati, social celebrities and artists opened. Wu Hufan wrote the Chinese signboard. The literary quotation came from the famous poem Li Bai used to describe Yang Kweifei, which was " the cloud reminds me of her clothing, the peony reminds me of her face", symbolizing the company's vision to transform fashion and beautify women with clothes.

1. 文化沙龙

云裳是一次时装实业的合作，更是一场文化沙龙。草创之初，由留法雕塑家江小鹣担任美术设计，文化名人徐志摩、周瘦鹃、胡适、张禹九等均为股东，社交名媛唐瑛为宣传推广不遗余力，陆小曼也参与日常事务。《上海画报》《北洋画报》都对开幕当日的情景作了大篇报道，稍后包天笑又作了《到云裳去》一文，详细介绍公司情况，刊发于《晶报》上，是一时盛事。

《上海画报》于1925年在上海创刊，是上海画报界的领潮者，拉开了20世纪二三十年代中国画报"黄金时代"的序幕。关注都市女性是《上海画报》的一大特色，作为"新女性"典范的陆小曼多次登上《上海画报》。《北洋画报》于1927年在天津创刊，在当时的传媒界有"北方巨擘"的美誉。《上海画报》和《北洋画报》都刊登了云裳公司开幕式照片，可见云裳影响力之大。开幕典礼上，股东之一朱润生七岁的女儿穿着旗袍和皮鞋，一身摩登打扮，主持了仪式。陆小曼穿着素雅的旗袍同徐志摩一同出席开幕典礼，报刊称二人为"云裳公司发起人"。

这件女童旗袍袍身修长，窄袖高领，采用常见的丝绸质地，大花纹样，与成年女性的旗袍一样摩登。

1. Cultural Salon

Yungzong is not only cooperation of the fashion and industry, but also a cultural salon. At the initial stage, the graphic designer of the company was Jiang Xiaojian, a sculptor who studied in France. Cultural celebrities such as Xu Zhimo, Zhou Shoujuan, Hu Shi and Zhang Yujiu were all shareholders of the company. Social celebrity Tang Ying enthusiastically promoted the company, and Lu Xiaoman was also involved in its operation. On the opening day, *Pictorial Shanghai* and *The Peiyang Pictorial News* both gave massive coverage of the scene, and later, Bao Tianxiao wrote an in-depth article about the company, published on *The Crystal*, creating a major event during that time.

Pictorial Shanghai was founded in Shanghai, 1925, and was the pioneer of Shanghai pictorial industry. It opened the prologue of the "golden age" for Chinese pictorial in 1920s-1930s. Focusing on urban women was a major feature of *Pictorial Shanghai*. As a model of "new women", Lu Xiaoman appeared in *Pictorial Shanghai* several times. *The Peiyang Pictorial News* was established in Tianjing in 1927, and had a reputation of "northern titan". Both of the papers published the opening ceremony proved the influence of Yungzong. On the ceremony, the seven-year-daughter of Zhu Runsheng, one of the shareholders, was wearing qipao and leather shoes, while hosting the ceremony. Lu Xiaoman, in a simple but elegant qipao, attended the ceremony along with Xu Zhimo. They were called founder of Yungzong by the newspapers.

This girl qipao is slim and with stand-up collar, using silk with floral pattern as material. It is as modern as adult qipao.

紫色提花真丝长袖夹旗袍
约1930-1940年代

领：立领，高7厘米
衣袖展长：99厘米
衣长：101厘米
胸宽：33厘米
腰宽：31厘米
摆宽：39厘米
侧开衩：24厘米
扣：盘扣12对，分布在领口3对，斜襟1对，侧门襟8对

Purple silk lined qipao with long sleeves and woven pattern
Around 1930s-1940s

Collar: Stand-up Collar, 7cm
Length of both sleeves: 99cm
Length: 101cm
Chest width: 33cm
Waist: 31cm
Length of hemline: 39cm
Side slits: 24cm
Button: 12 pairs of frog closures, 3 on collar, 1 on slant opening, 8 on the side

唐瑛
Tang Ying

唐瑛是中华医学会创办人唐乃安之女，也是活跃于上海社交舞台的名媛，与陆小曼并称“南唐北陆”。唐瑛是云裳公司的股东之一，并与徐志摩一起担任特别顾问。照片中她身穿格纹无袖旗袍，双手插在黑色皮毛袖笼里，脚着黑色高跟皮鞋，现代感十足，而挽起的发髻又为她增添几分古典气息。

Tang Ying is the daughter of the founder of Tang Naian, who was the Chinese Medical Association. And she was a socialite actived on the Shanghai social stage. Along with Lu Xiaoman, they were called “South Tang North Lu”. Tang Ying is one of Yongzong’s shareholders, and was also a special consultant for the company. In the picture she was wearing plaid sleeveless qipao, hands in black fur glove-like bags, foots in black heels. Despite the modern look, her hair bun still showed her traditional beauty.

黑地格纹无袖单旗袍
约1930-1940年代

领：立领，高2.5厘米
肩宽：43厘米
衣长：114厘米
胸宽：39厘米
腰宽：36厘米
摆宽：43厘米
侧开衩：25厘米
扣：盘花扣2对，分布在领口1对，斜襟1对；盘扣1对，在侧门襟；揿纽7对

Black plaid sleeveless qipao
Around 1930s-1940s

Collar: Stand-up Collar, 2.5cm
Shoulder: 43cm
Length: 114cm
Chest width: 39cm
Waist: 36cm
Length of hemline: 43cm
Side slits: 25cm
Button: 2 pairs of floral frog closures, 1 on collar, 1 on slant opening; 1 pair of frog closures on the side; 7 metal press studs

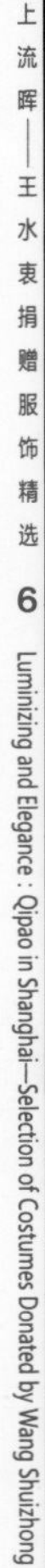

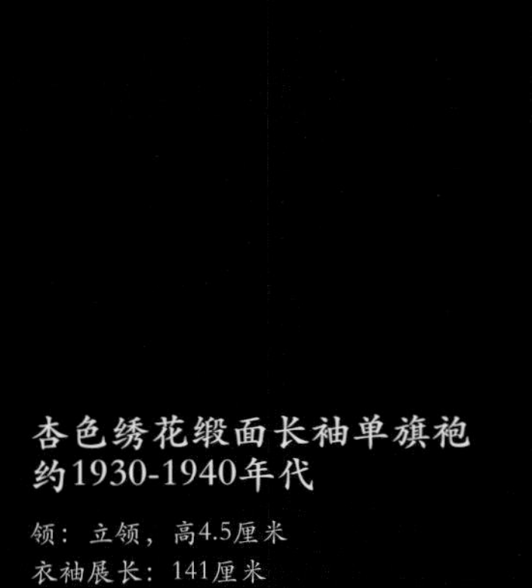

杏色绣花缎面长袖单旗袍
约1930-1940年代

领：立领，高4.5厘米
衣袖展长：141厘米
衣长：128厘米
胸宽：40厘米
腰宽：38厘米
摆宽：46厘米
侧开衩：39厘米
扣：盘香扣2对，分布在领口1对，
斜襟1对；盘扣9对，分布在领口1对，
侧门襟1对；揿纽1对

Apricot damask long-sleeved qipao with embroidery decoration
Around 1930s-1940s

Collar: Stand-up Collar, 4.5cm
Length of both sleeves: 141cm
Length: 128cm
Chest width: 40cm
Waist: 38cm
Length of hemline: 46cm
Side slits: 39cm
Button: 2 pairs of incense coil frog closures,
1 on collar, 1 on slant opening; 9 pairs of frog closures,
1 on collar, 1 on the side; 1 metal press stud

杏色绣花缎面长袖单旗袍
约1930-1940年代

领：立领，高2.5厘米
衣袖展长：114.5厘米
衣长：109厘米
胸宽：41厘米
腰宽：37厘米
摆宽：45厘米
侧开衩：12厘米
扣：盘扣3对，分布在领口1对、斜襟1对，侧门襟1对；揿纽6对

Apricot damask long-sleeved qipao with embroidery decoration
Around 1930s-1940s

Collar: Stand-up Collar, 2.5cm
Length of both sleeves: 114.5cm
Length: 109cm
Chest width: 41cm
Waist: 37cm
Length of hemline: 45cm
Side slits: 12cm
Button: 3 pairs of frog closures, 1 on collar, 1 on slant opening, 1 on the side; 6 metal press studs

陆小曼
Lu Xiaoman

陆小曼是集传统与现代于一身的女子，她喜爱旧戏，擅长工笔，但同时思想开放，忠于自我。陆小曼喜爱旗袍，在云裳服装公司的开幕典礼上，作为股东的她穿着旗袍出席，也算为公司所主推的服装品类作了代言。

Lu Xiaoman was a woman who perfectly balanced old and new. She loved old plays and was good at fine brushwork, but also open-minded and loyaled to herself. Lu Xiaoman loved qipao and was wearing one when attending the opening ceremony of Yungzong. This could count for her endorsement for the clothing.

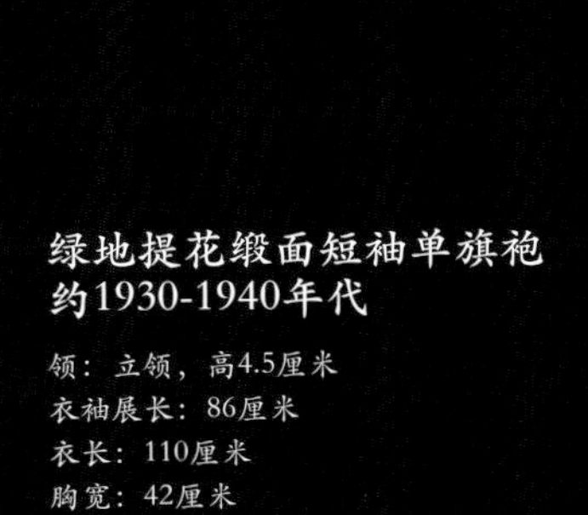

绿地提花缎面短袖单旗袍
约1930-1940年代

领：立领，高4.5厘米
衣袖展长：86厘米
衣长：110厘米
胸宽：42厘米
腰宽：38厘米
摆宽：44厘米
侧开衩：14厘米
扣：揿纽9对

Green damask short-sleeved qipao with woven pattern
Around 1930s-1940s

Collar: Stand-up Collar, 4.5cm
Length of both sleeves: 86cm
Length: 110cm
Chest width: 42cm
Waist: 38cm
Length of hemline: 44cm
Side slits: 14cm
Button: 9 metal press studs

张幼仪
Zhang Youyi

云裳公司股东众多，张幼仪的弟弟张禹九是其中之一。梁实秋、刘英士和张禹九的孙女张邦梅先后撰文讲述云裳实际是张幼仪女士所经营的往事。

我跪在爸妈家的桃花心木箱前，紧握着幼仪每天穿的黑旗袍，仿佛那衣衫可以召唤我姑婆似的。嵌织着莲花纹样的平滑织物，看起来仿如星光点点的黑色池塘。

——张邦梅《小脚与西服》

The Yungzong Company had many shareholders. Zhang Yujiu, younger brother of Zhang Youyi, was one of them. Liang Shiqiu , Liu Yingshi and Zhang Bangmei, who was the grand-daughter of Zhang Yujiu, used to write the stories about Yungzong. These stories told that Yungzong was run by Zhang Youyi in fact.

I knelt at my parents' home in front of the mahogany box, holding the black qipao worn by her every day, as if the dress could call my aunt. The smooth fabric woven with lotus patterns looks like a starry black pond.

—— *Bound feet and Suits by Zhang Bangmei*

黑色轧花丝绒短袖单旗袍
约1930-1940年代

领：立领，高3厘米
衣袖展长：47厘米
衣长：120厘米
胸宽：41厘米
腰宽：39厘米
摆宽：49厘米
侧开衩：31厘米
扣：盘香扣2对，分布在领口1对，斜襟1对；盘扣8对，在侧门襟；揿纽2对

Black velvet short-sleeved qipao
Around 1930s-1940s

Collar: Stand-up Collar, 3cm
Length of both sleeves: 47cm
Length: 120cm
Chest width: 41cm
Waist: 39cm
Length of hemline: 49cm
Side slits: 31cm
Button: 2 pairs of incense coil frog closures, 1 on collar, 1 on slant opening;
8 pairs of frog closures on the side;
2 metal press studs

2. 文化品牌

云裳公司的经营策略，是以世界最流行的装束参以中国人的习惯，以国货为主辅以洋货制作时装，定价低廉以便普及。公司的广告充满了文艺气息，以著名画家吴湖帆的篆书题写店名，用祥云衬托莲花的图案作为店标。务求打造中国人自己的服装、艺术、文化品牌，1927年8月6日云裳在《申报》刊登开业广告，用文艺的语言书写道:

要穿最漂亮的衣服
到云裳去
要想最有意识的衣服
到云裳去
要想最精美的打扮
到云裳去
要个性最分明的式样
到云裳去

2. Artistic Brand

The management strategy of Yungzong Company was to fit the most popular costume in the world into the Chinese custom, using mainly the domestic goods supplemented with foreign goods to make the fashion, with low price so as to popularize. The company's advertisements were full of artistic atmosphere, with the famous painter Wu Hufan inscribing the name of the shop with the seal character, and using the lotus pattern with Chinese traditional cloud as the shop logo. Committed to build our own clothing, art and cultural brands, Yungzong advertised the opening ceremony on *Sun Pao* as below,

Wear the most beautiful clothes
Go to the Yungzong
Think of the most conscious clothes
Go to the Yungzong
Think of the most delicate apparel
Go to the Yungzong
Want the most unique style
Go to the Yungzong

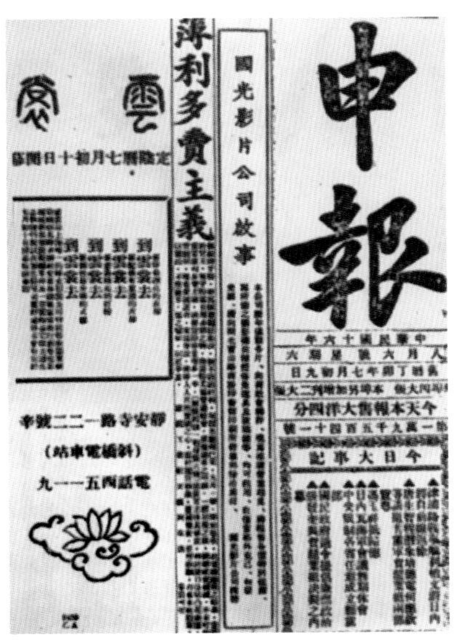

云裳公司开业广告
The opening commercial of Yungzong

云裳公司广告（载《旅行杂志》1927年冬季号）
Advertisement of Yungzong
(winter issue, 1927, China Traveler)

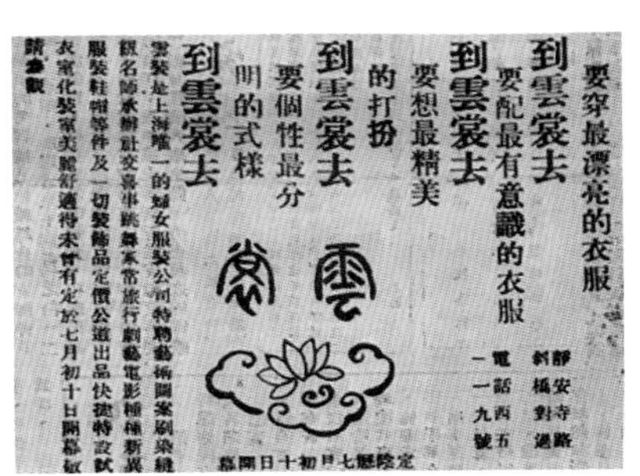

云裳公司广告（载《上海画报》第262期，1927年8月12日）
Advertisement of Yungzong
(issue 262nd, August 12th 1927, the Pictorial shanghai)

云裳公司广告（载《上海漫画》1928年第79期）
Advertisement of Yungzong
(issue 79th, 1928, the Shanghai Sketch)

3. 时尚秀场

上海是中国最早举办时装秀的城市，时装秀的发展与时装的发展是同步的，旗袍的展示是时装秀上必不可少的主题。最初的时装秀往往穿插于游艺活动或展览会之中，举办时装表演的目的是“寓美术教育于游戏”和“表现服装料作之如何可以充分利用”。这种美术目的的时装表演在当时深受文化界人士追捧，以“空前之美术服装公司”为标榜的云裳服装公司就很热衷于参与这样的活动。

3. Fashion Show

Shanghai is the earliest city in China to have fashion show. The development of fashion and fashion show are synchronous, and qipao was the inevitable theme on fashion show. Fashion show first appeared on recreation fairs and exhibition, with the purpose of artistic education and showing how to make full use of fabric. This kind of fashion show, which had artistic purposes, was very popular in the cultural circle at that time. The clothing company, which named itself “the unprecedented art clothing company” was keen to take part in such activities.

1927年云裳公司时装表演（载《上海画报》1927年第292期）
1927 Yungzong fashion show (issue 292nd, 1927. The Pictorial shanghai)

1927年10月云裳公司加入了汽车展览会，表演新装展示，根据现存图像资料显示，所展示的服装均为旗袍。

The Yungzong Clothing Company jointed the Automobile Expo, showing new collection. According to the picture survived, all the collection was qipao.

1930年云裳公司时装表演（载《良友》1930年第52期）
1930 Yungzong fashion show (issue 52nd, 1930. The Young Companion)

1930年国货运动大会的时装表演中有两场旗袍秀，一场为短款旗袍展示，一场为长款旗袍展示，展示中模特穿着旗袍均为云裳公司提供。

1930 China-made Production Expo had two qipao fashion shows, one of them was short qipao fashion show and the other was long qipao fashion show. The qipao worn by the models in the picture were supplied by Yungzong.

绿地提花镶花边真丝
倒大袖夹旗袍
约1930年代

领：立领，高5.5厘米
衣袖展长：100厘米
衣长：115厘米
胸宽：40厘米
腰宽：40厘米
摆宽：61厘米
扣：盘扣4对，分布在侧门襟；
揿纽6对

Green bell-shaped sleeves
silk qipao bound with lace
Around 1930s

Collar: Stand-up Collar, 5.5cm
Length of both sleeves: 100cm
Length: 115cm
Chest width: 40cm
Waist: 40cm
Length of hemline: 61cm
Button: 2 pairs of frog closures on the side;
6 metal press studs

粉色镶边缎面中袖单旗袍
约1930年代

领：立领，高5厘米
衣袖展长：65厘米
衣长：106厘米
胸宽：36厘米
腰宽：35厘米
摆宽：45厘米
侧开衩：13厘米
扣：盘扣11对，分布在领口2对，斜襟1对，侧门襟8对

Pink damask half-sleeved qipao with trimming
Around 1930s

Collar: Stand-up Collar, 5cm
Length of both sleeves: 65cm
Length: 106cm
Chest width: 36cm
Waist: 35cm
Length of hemline: 45cm
Side slits: 13cm
Button: 11 pairs of frog closures, 2 on collar, 1 on slant opening, 8 on the side

提花镶花边缎面长袖夹旗袍
约1930年代

领：立领，高5厘米
衣袖展长：110厘米
衣长：116厘米
胸宽：39厘米
腰宽：39厘米
摆宽：61厘米
扣：盘香扣6对，分布在领口1对，
斜襟1对，侧门襟4对

Damask long-sleeved qipao with lining and woven pattern bound with lace
Around 1930s

Collar: Stand-up Collar, 5cm
Length of both sleeves: 110cm
Length: 116cm
Chest width: 39cm
Waist: 39cm
Length of hemline: 61cm
Button: 6 pairs of incense coil frog closures, 1 on collar, 1 on slant opening, 4 on the side

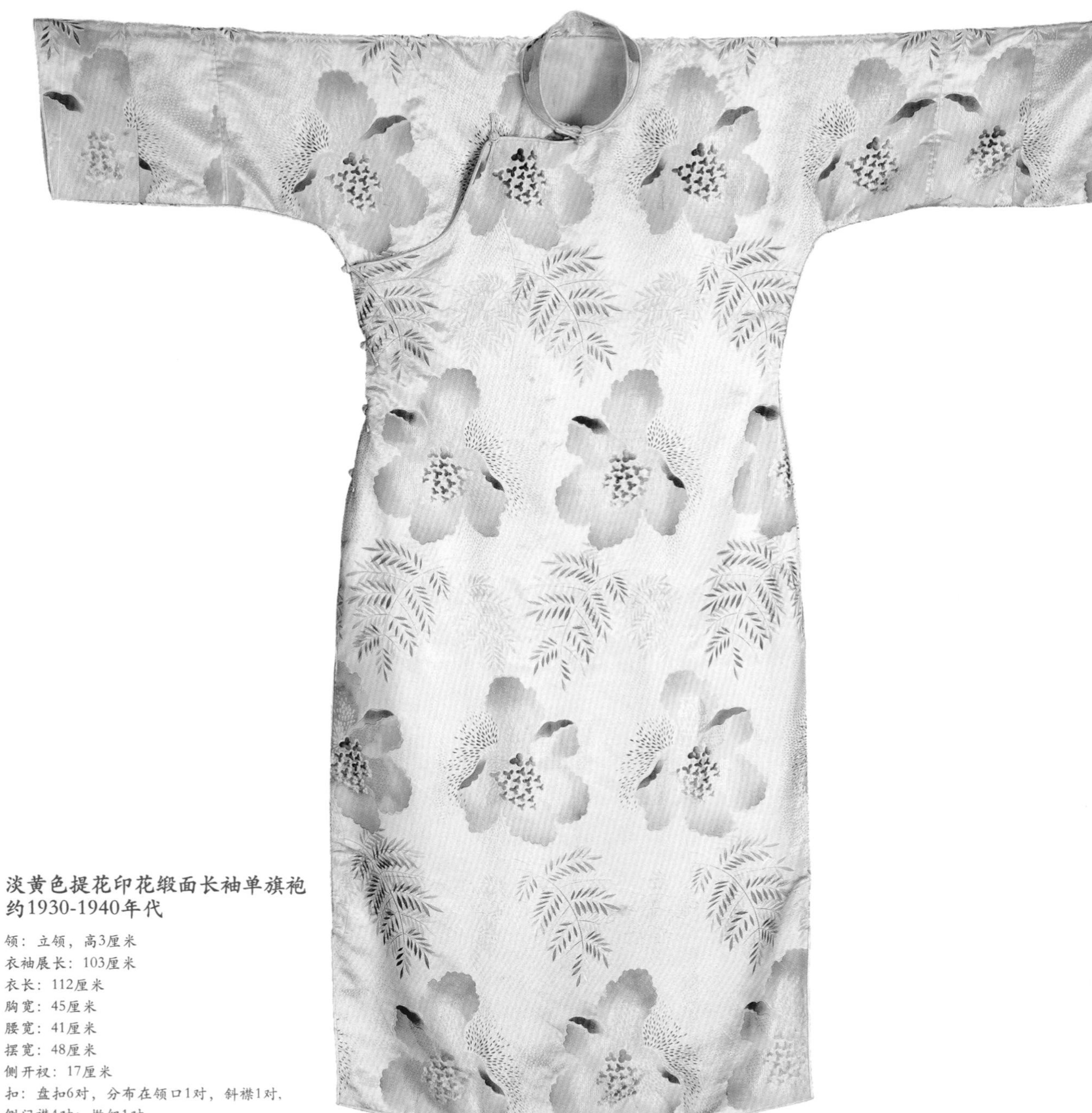

淡黄色提花印花缎面长袖单旗袍
约1930-1940年代

领：立领，高3厘米
衣袖展长：103厘米
衣长：112厘米
胸宽：45厘米
腰宽：41厘米
摆宽：48厘米
侧开衩：17厘米
扣：盘扣6对，分布在领口1对，斜襟1对，侧门襟4对；揿纽1对

Pale yellow damask long-sleeved qipao with printed and woven pattern
Around 1930s-1940s

Collar: Stand-up Collar, 3cm
Length of both sleeves: 103cm
Length: 112cm
Chest width: 45cm
Waist: 41cm
Length of hemline: 48cm
Side slits: 17cm
Button: 2 pairs of frog closures, 1 on collar, 1 on slant opening, 4 on the side; 1 metal press stud

蓝色蕾丝中袖单旗袍
约1920-1930年代

领：立领，高6厘米
衣袖展长：95厘米
衣长：107厘米
胸宽：39厘米
腰宽：38厘米
摆宽：54厘米
侧开衩：7厘米
扣：盘香扣9对，分布在领口2对，
斜襟1对，侧门襟6对

Blue half-sleeved lace qipao
Around 1920s-1930s

Collar: Stand-up Collar, 6cm
Length of both sleeves: 95cm
Length: 107cm
Chest width: 39cm
Waist: 38cm
Length of hemline: 54 cm
Side slits: 7cm
Button: 2 pairs of incense coil frog closures,
2 on collar, 1 on slant opening,
6 on the side

紫色提花镶花边真丝长袖单旗袍
约1920-1930年代

领：立领，高2厘米
衣袖展长：109厘米
衣长：108厘米
胸宽：45厘米
腰宽：42厘米
摆宽：51厘米
侧开衩：20厘米
扣：盘扣8对，分布在领口1对、斜襟1对，侧门襟6对

Purple silk long-sleeved qipao bound with lace
Around 1920s-1930s

Collar: Stand-up Collar, 2cm
Length of both sleeves: 109cm
Length: 108cm
Chest width: 45cm
Waist: 42cm
Length of hemline: 51cm
Side slits: 20cm
Button: 8 pairs of frog closures, 1 on collar, 1 on slant opening, 6 on the side

4、先锋设计

1930年代的时装是美术家时代的时装，相较于巴黎、纽约这些的西方大都市，彼时上海时装最具中国特色的核心即是美术。时装从设计到制成，往往由设计师在报刊上刊登设计款式，待读者看中制为成衣。云裳时装公司是其中的先锋，草创之初担任服装设计的是留法雕塑家江小鹣，1928年江小鹣卸任后，公司聘请了当时的画坛新秀，后来成为著名漫画大师的叶浅予担任设计师。

4. Avant-garde Design

The fashion in 1930s was the fashion of the artists. Compared with Paris and New York, the main Chinese character during that time was art. From design to manufacture, fashion designers often published designs in newspapers and magazines, waiting for readers' selection to make clothes. The graphic designer of the company was first Jiang Xiaojian, a sculptor who studied in France, then became the well-known painter Ye Qianyu.

江小鹣
Jiang Xiaojian

江小鹣早年留学日本，学习素描和油画，回国后供职于上海美专。后又留学法国，学习雕塑，回国后在虹桥路建立雕塑工作室。1928年，江小鹣与张辰伯发起组织艺苑绘画研究所。

Jiang Xiaojian was in Japan to study sketches and oil paintings in his early years. He returned to China to work in Shanghai Academy of art. After that he went to study in France, learning sculpture, and set up a sculpture studio on Hongqiao Road after he returned. In the year of 1928, Jiang Xiaojian initiated the Research Institute of art painting.

丁聪画叶浅予像
Portrait of Ye Qianyu by Ding Cong

叶浅予在自传《细叙沧桑记流年》中写道："我对美国同一出版社出版的《Vogue》时装杂志也发生兴趣，开始在《上海漫画》和《图画晨报》发表时装设计图，还被一家（云裳）时装公司聘为设计师。"

Ye Qianyu wrote in his autobiography, "I am also interested in *Vogue* magazine, starting to publish fashion design sketch on *The Shanghai Sketch* and *The Chen Pao Miscellany*, and hired as designer by a fashion company (Yungzong).

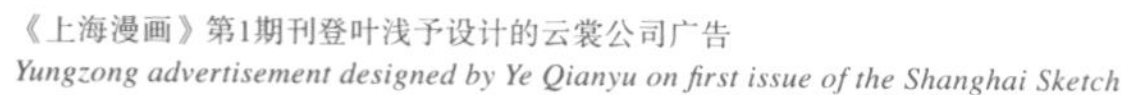
《上海漫画》第1期刊登叶浅予设计的云裳公司广告
Yungzong advertisement designed by Ye Qianyu on first issue of the Shanghai Sketch

叶浅予《最流行之新装》
"The New Popular Style" by Ye Qianyu

叶浅予在1931年至1933年期间为《玲珑》画服装设计插画五十余幅，其中旗袍和带有旗袍元素的服装占了大多数。此设计图刊登于1931年第18期《玲珑》，图中妇女卷发，穿着格纹短袖旗袍，长度至小腿中部，开衩及膝。

Ye Qianyu drew more than 50 sketches for the *Linloon Magazine* during the years 1930 to 1933, among which the majority were qipao and clothing with qipao elements. This design was published on the *Linloon Magazine* 18th issue, 1931. The woman in the picture was curly, wearing short-sleeved plaid qipao, the hemline of which went up to the mid-calf and with split up to the knee.

叶浅予《冬季大衣的时装》
"Winter outfit with coat" by Ye Qianyu

这幅皮大衣的设计图刊登于1932年第43期《玲珑》。大衣内搭均为长至脚踝的旗袍，一款为浅色的镶边旗袍，在浅色大衣的领口和摆下露出镶边部分，显得错落有致；一款为深色圆点或小碎花旗袍，搭配深色带毛领大衣，衣襟敞开，露出内里的旗袍，隐隐绰绰。

This design of fur coat was published on the *Linloon Magazine*, issue 43rd 1932. Inside the coat are both long qipao with hemline to the foot. One of them is light-colored with decoration band, exposed at the cuffs and hemline beneath the light-colored coat; another one is dark dots or floral qipao dressed with dark fur coat with a fur collar, half opened to show the qipao inside.

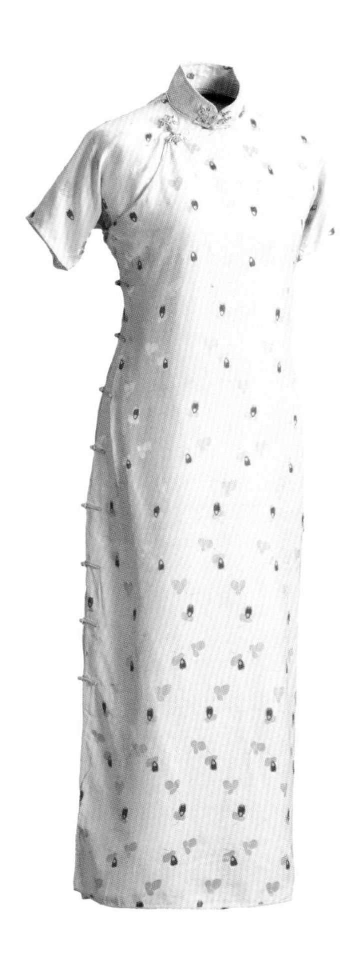

藕色提花真丝中袖单旗袍
约1930-1940年代

领：立领，高4厘米
衣衣袖展长：64厘米
衣长：125厘米
胸宽：42厘米
腰宽：41厘米
摆宽：50厘米
侧开衩：31厘米
扣：盘花扣2对，分布在领口1对，斜襟1对；
盘扣10对，分布在领口1对，侧门襟9对

Beige silk half-sleeved qipao with woven pattern
Around 1930s-1940s

Collar: Stand-up Collar, 4cm
Length of both sleeves: 64cm
Length: 125cm
Chest width: 42cm
Waist: 41cm
Length of hemline: 50cm
Side slits: 31cm
Button: 2 pairs of floral frog closures,
1 on collar, 1 on slant opening;
10 pairs of frog closures,
1 on the collar,
9 on the side

藏青色衬骆驼绒提花真丝短袖夹旗袍
约1930-1940年代

领：立领，高7厘米
衣袖展长：80厘米
衣长：124厘米
胸宽：39厘米
腰宽：36厘米
摆宽：46厘米
侧开衩：30厘米
扣：盘香扣12对，分布在领口4对，
斜襟1对，侧门襟7对；揿纽4对

Navy blue short-sleeved silk qipao with camel hair cloth as lining
Around 1930s-1940s

Collar: Stand-up Collar, 7cm
Length of both sleeves: 80cm
Length: 124cm
Chest width: 39cm
Waist: 36cm
Length of hemline: 46cm
Side slits: 30cm
Button: 12 pairs of incense coil frog closures,
4 on collar, 1 on slant opening, 7 on the side;
4 metal press studs

三、杂志中的时尚话题

Three
Fashion Topics in Magazines

三、杂志中的时尚话题

上海不仅是时尚的中心，也是传媒的中心。民国时期上海滩上一些著名的报刊，从内容和功能上来看，与今日的时尚杂志无异。女性的妆容和服饰是这些时尚杂志关注的话题之一，旗袍时装则经常出现在时尚杂志的插画和专题中。

1、《良友》和《妇人画报》

创刊于1926年的《良友》是大型生活类画报，女性服饰妆容和摩登女性名人始终是它最受欢迎的专题之一。《妇人画报》创刊于1933年，是良友公司旗下的妇女生活杂志，它关注都市女性的日常生活，穿着打扮，两性问题，此外还开设了小说专栏。

1940年的《良友》画报刊登过一期名为“旗袍的旋律”的专题，对旗袍的发展做了梳理。

Three: Fashion Topics in Magazines

Shanghai was not only the center of fashion, but also the center of media. In the Republic of China, some famous newspapers and periodicals in Shanghai were no different from today's fashion magazines in terms of content and function. Make-up and fashion were always hot topics for these magazines, while qipao often appeared in illustrations or in the special topic.

1. *The Young Companion* and *The Women's Pictorial*

Founded in 1926, *The young companion* was a large life style pictorial; women's makeup and modern female celebrities have always been one of its most popular topics. *The Women's Pictorial* was established in 1933, is the life style magazine under the Young Companion Company. It focused on the daily life of urban women, dressing and gender topics. It also had novel column.

On *The Young Companion* published in 1940, one of the issues had an in-depth report called "The Melody of Qipao", interpreted the progress of qipao.

《旗袍的旋律》
The Melody of Qipao

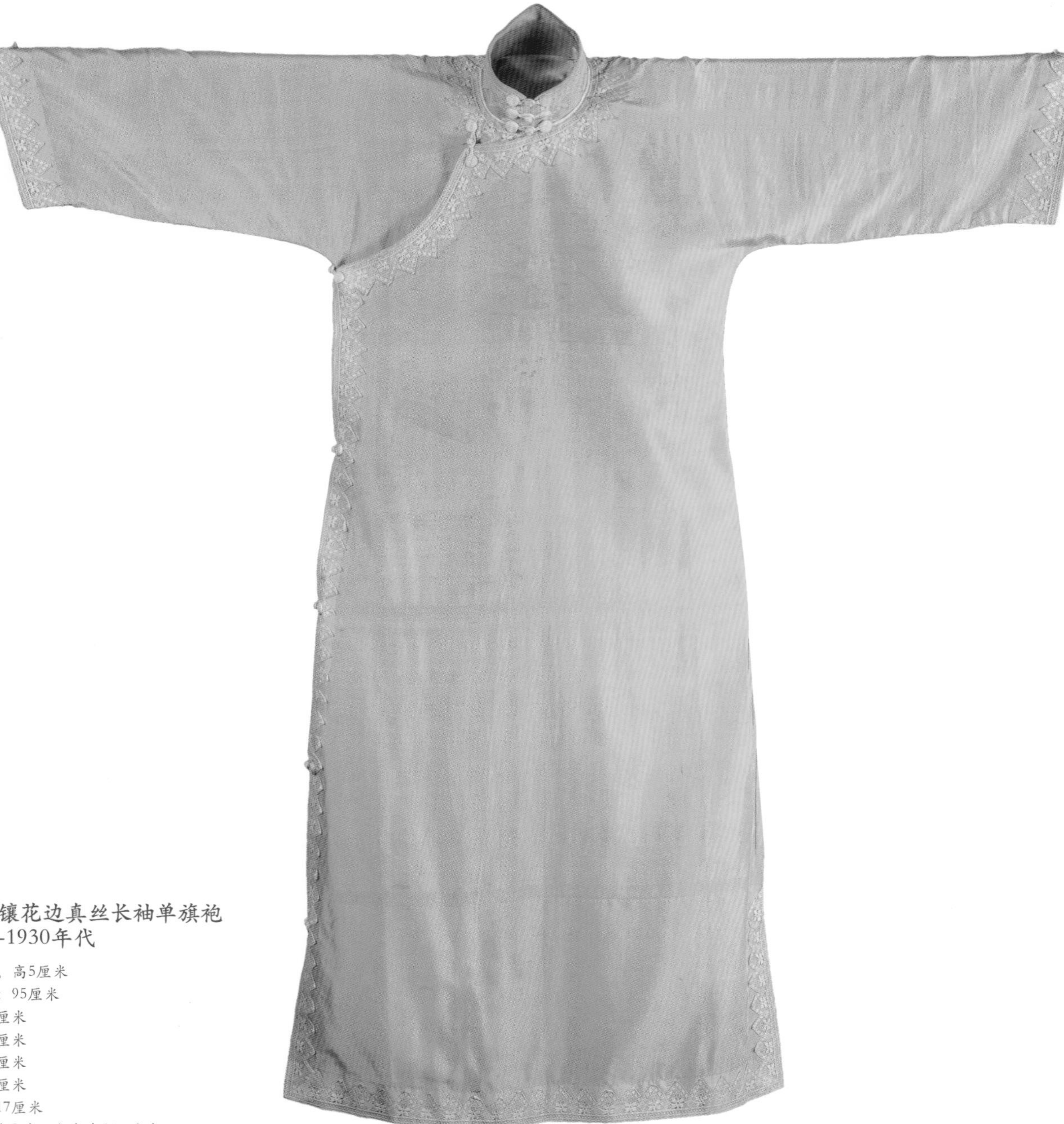

杏红色镶花边真丝长袖单旗袍
约1920-1930年代

领：立领，高5厘米
衣袖展长：95厘米
衣长：97厘米
胸宽：36厘米
腰宽：36厘米
摆宽：47厘米
侧开衩：17厘米
扣：盘香扣9对，分布在领口3对，
斜襟1对，侧门襟5对

Apricot long-sleeved silk qipao bound with lace
Around 1920s-1930s

Collar: Stand-up Collar, 5cm
Length of both sleeves: 95cm
Length: 97cm
Chest width: 36cm
Waist: 36cm
Length of hemline: 47cm
Side slits: 17cm
Button: 9 pairs of incense coil frog closures,
3 on collar, 1 on slant opening,
5 on the side

淡紫色提花真丝中袖单旗袍
约1930-1940年代

领：立领，高6厘米
衣袖展长：93厘米
衣长：127厘米
胸宽：36厘米
腰宽：36厘米
摆宽：50厘米
侧开衩：34厘米
扣：盘扣13对，分布在领口5对

Light purple half-sleeved silk qipao with woven pattern
Around 1930s-1940s

Collar: Stand-up Collar, 6cm
Length of both sleeves: 93cm
Length: 127cm
Chest width: 36cm
Waist: 36cm
Length of hemline: 50cm
Side slits: 34cm
Button: 13 pairs of frog closures, 5 on collar

黑地提花八字襟无袖单旗袍
约1940年代

领：立领，高4厘米
肩宽：43厘米
衣长：102厘米
胸宽：39厘米
腰宽：35.5厘米
摆宽：43厘米
侧开衩：9厘米
扣：盘香扣1对，在侧门襟；揿纽11对

Black sleeveless qipao with symmetrical opening
Around 1940s

Collar: Stand-up Collar, 4cm
Shoulder: 43cm
Length: 102cm
Chest width: 39cm
Waist: 35.5cm
Length of hemline: 43cm
Side slits: 9cm
Button: 1 pair of incense coil frog closures on the side;
11 metal press studs

紫色蕾丝八字襟无袖单旗袍
约1940年代

领：立领，高2.5厘米
肩宽：40厘米
衣长：102厘米
胸宽：38厘米
腰宽：37厘米
摆宽：46厘米
侧开衩：19厘米
扣：盘扣4对，分布在领口1对，
斜襟1对，侧门襟2对；揿纽1对

Purple lace sleeveless qipao with symmetrical opening
Around 1940s

Collar: Stand-up Collar, 2.5cm
Shoulder: 40cm
Length: 102cm
Chest width: 38cm
Waist: 37cm
Length of hemline: 46cm
Side slits: 19cm
Button: 4 pairs of frog closures, 1 on collar,
1 on slant opening, 2 on the side;
1 metal press studs

方雪鹄《秋季新装》
Winter Outfit by Fang Xuehu

方雪鹄是上海美专科班出身，西洋画功底很好，在画时装设计图时也往往采用素描、水彩等画法，风格十分鲜明。1930年总第50期《良友》刊登方雪鹄《秋季新装》，运用淡雅的颜色设计了几款旗袍。

Fang Xuehu was graduated from Shanghai Academy of Art. He was good at western painting and often used sketching, watercolor and other painting methods when drawing fashion design sketches. *The Young Companion*, issue 50th 1930 published *Winter Outfit* by Fang Xuehu, using light colors to design several qipao.

方雪鹄《新装》
The New Dress by Fang Xuehu

1937年第47期《妇人画报》刊登方雪鹄所设计的《新装》，即以素描的方式呈现他的设计理念，图中女性穿着大花图案的短袖镶边旗袍。

The New Dress published on issue 47th, 1937 of *The Women's Pictorial*, using sketches to present his ideas of design. The women in the picture were wearing short-sleeved qipao with decorating trimmings.

蓝地印花真丝中袖单旗袍
约1930年代

领：立领，高6.5厘米
衣袖展长：96厘米
衣长：107厘米
胸宽：39厘米
腰宽：39厘米
摆宽：51厘米
侧开衩：8厘米对
扣：盘扣5对，分布在领口3对，斜襟1对，侧门襟1对；揿纽3对

Blue ground half-sleeved silk qipao with printed patten
Around 1930s

Collar: Stand-up Collar, 6.5cm
Length of both sleeves: 9.5cm
Length: 107cm
Chest width: 39cm
Length of hemline: 51cm
Side slits: 8cm
Button: 5 pairs of frog closures, 3 on collar, 1 on slant opening, 1 on the side; 3 metal press studs

2.《玲珑》

《玲珑》创刊于1931年3月18日，每周三出版一期。在第一期中即明确表示杂志的目标为“增进‘妇女’优美生活，提倡社会高尚‘娱乐’”。《玲珑》关注女性的妆容和服饰，经常发布欧洲的流行资讯，也关注女性的生活方式和进步思想，探讨“什么是真正的摩登”的话题。

2. *Linloon Magazine*

Linloon Magazine was founded on March 18th, 1931, issued every Wednesday. On the first issue, the magazine stated that its aim was “to promote the beautiful life of women and the noble entertainment of society”. *Linloon Magazine* focused on women’s make-up and costume, often released popular information in Europe, but also focused on women’s lifestyle and the progress of thinking, to discuss their topic of “what is true modern”.

巴黎婦女之長短競爭

（何美娟）

近日巴黎女界。對于頭髮式樣及衣裝。曾有一度之劇烈競爭。一般年輕姑娘。皆主張短髮垂肩。高裙露膝。而年紀稍長之少婦。却主張雲髮蓬鬆。長裙曳地。一時競爭頗烈。筆刀舌劍。哄動一時。社會目光。咸集中於此種爭辯。最感煩難者。爲一般成衣匠。蓋雙方主張各異。此長彼短。不知誰是誰非。徘徊兩端。不知所從。

1932年第48期《玲珑》
Issue 48th 1932 Linloon Magazine

1932年第48期的《玲珑》则刊登了《巴黎妇女之长短竞争》一文，称巴黎女界“年轻姑娘，皆主张短发垂肩，高裙露膝。而年纪稍长之少妇，却主张云发蓬松，长裙曳地。一时竞争颇烈。”

Issue 48th 1932 *Linloon Magazine* published “The Long and Short Competition of Women in Paris” which says that in Paris, “ Young ladies prefer short hair to the shoulder and short skirt to expose the knees; while more matured women prefers fluffy hair and long dress. The competition was fierce.”

淡紫色提花真丝短袖单旗袍
约1930-1940年代

领：立领，高4.5厘米
衣袖展长：62厘米
衣长：131厘米
胸宽：43厘米
腰宽：41厘米
摆宽：46厘米
侧开衩：34厘米
扣：盘花扣2对，分布在领口1对、斜襟1对；盘扣9对，分布在领口1对、侧门襟8对；揿纽3对

Light purple short-sleeved silk qipao with woven pattern
Around 1930s-1940s

Collar: Stand-up Collar, 4.5cm
Length of both sleeves: 62cm
Length: 131cm
Chest width: 43cm
Waist: 41cm
Length of hemline: 46cm
Side slits: 34cm
Button: 2 pairs of floral frog closures, 1 on collar, 1 on slant opening; 9 pairs of frog closures, 1 on collar, 8 on the side; 3 metal press studs

黑色提花真丝短袖单旗袍
约1930-1940年代

领：立领，高6.5厘米
衣袖展长：53厘米
衣长：132厘米
胸宽：39厘米
腰宽：37厘米
摆宽：44厘米
侧开衩：30厘米
扣：盘香扣12对，分布在领口3对，斜襟1对，侧门襟8对

Black silk short-sleeved qipao with woven pattern
Around 1930s-1940s

Collar: Stand-up Collar, 6.5cm
Length of both sleeves: 53cm
Length: 132cm
Chest width: 39cm
Waist: 37cm
Length of hemline: 44cm
Side slits: 30cm
Button: 12 pairs of incense coil frog closures, 3 on collar, 1 on slant opening, 8 on the side

黄色蕾丝无袖单旗袍
约1930-1940年代

领：立领，高5厘米
肩宽：42.5厘米
衣长：133.5厘米
胸宽：40厘米
腰宽：38厘米
摆宽：41厘米
侧开衩：38厘米
扣：盘花扣3对，分布在领口1对，斜襟1对，侧门襟1对；揿纽7对

Yellow sleeveless lace qipao
Around 1930s-1940s

Collar: Stand-up Collar, 5cm
Shoulder: 42.5cm
Length: 133.5cm
Chest width: 40cm
Waist: 38cm
Length of hemline: 48cm
Side slits: 38cm
Button: 3 pairs of floral frog closures, 1 on collar, 1 on slant opening, 1 on the side; 7 metal press studs

绿色印花纱短袖单旗袍
约1930-1940年代

领：立领，高6.5厘米
衣袖展长：58厘米
衣长：129厘米
胸宽：36厘米
腰宽：34厘米
摆宽：38厘米
侧开衩：48厘米
扣：盘扣17对，分布在领口4对，斜襟1对，侧门襟12对

Green yarn short-sleeved qipao with printed pattern
Around 1930s-1940s

Collar: Stand-up Collar, 6.5cm
Length of both sleeves: 58cm
Length: 129cm
Chest width: 36cm
Waist: 34cm
Length of hemline: 38cm
Side slits: 48cm
Button: 17 pairs of frog closures, 4 on collar, 1 on slant opening, 12 on the side

蓝色衬绒绣花真丝中袖夹旗袍
约1930-1940年代

领：立领，高6厘米
衣袖展长：92厘米
衣长：128厘米
胸宽：46厘米
腰宽：45厘米
摆宽：52厘米
侧开衩：28厘米
扣：盘花扣11对，分布在领口2对，
斜襟1对，侧门襟8对；揿纽2对

Blue embroidery half-sleeved silk qipao with velvet lining
Around 1930s-1940s

Collar: Stand-up Collar, 6cm
Length of both sleeves: 92cm
Length: 128cm
Chest width: 46cm
Waist: 45cm
Length of hemline: 52cm
Side slits: 28cm
Button: 11 pairs of floral frog closures,
2 on collar, 1 on slant opening, 8 on the side;
2 metal press studs

咖啡色提花真丝中袖单旗袍
约1930-1940年代

领：立领，高5.5厘米
衣袖展长：87厘米
衣长：124厘米
胸宽：42厘米
腰宽：52厘米
摆宽：46厘米
侧开衩：25厘米
扣：盘花扣11对，分布在领口2对，
斜襟1对，侧门襟8对

Dark coffee half-sleeved qipao with woven pattern
Around 1930s-1940s

Collar: Stand-up Collar, 5.5cm
Length of both sleeves: 87cm
Length: 124cm
Chest width: 42cm
Waist: 52cm
Length of hemline: 46cm
Side slits: 25cm
Button: 11 pairs of floral frog closures,
2 on collar, 1 on slant opening,
8 on the side

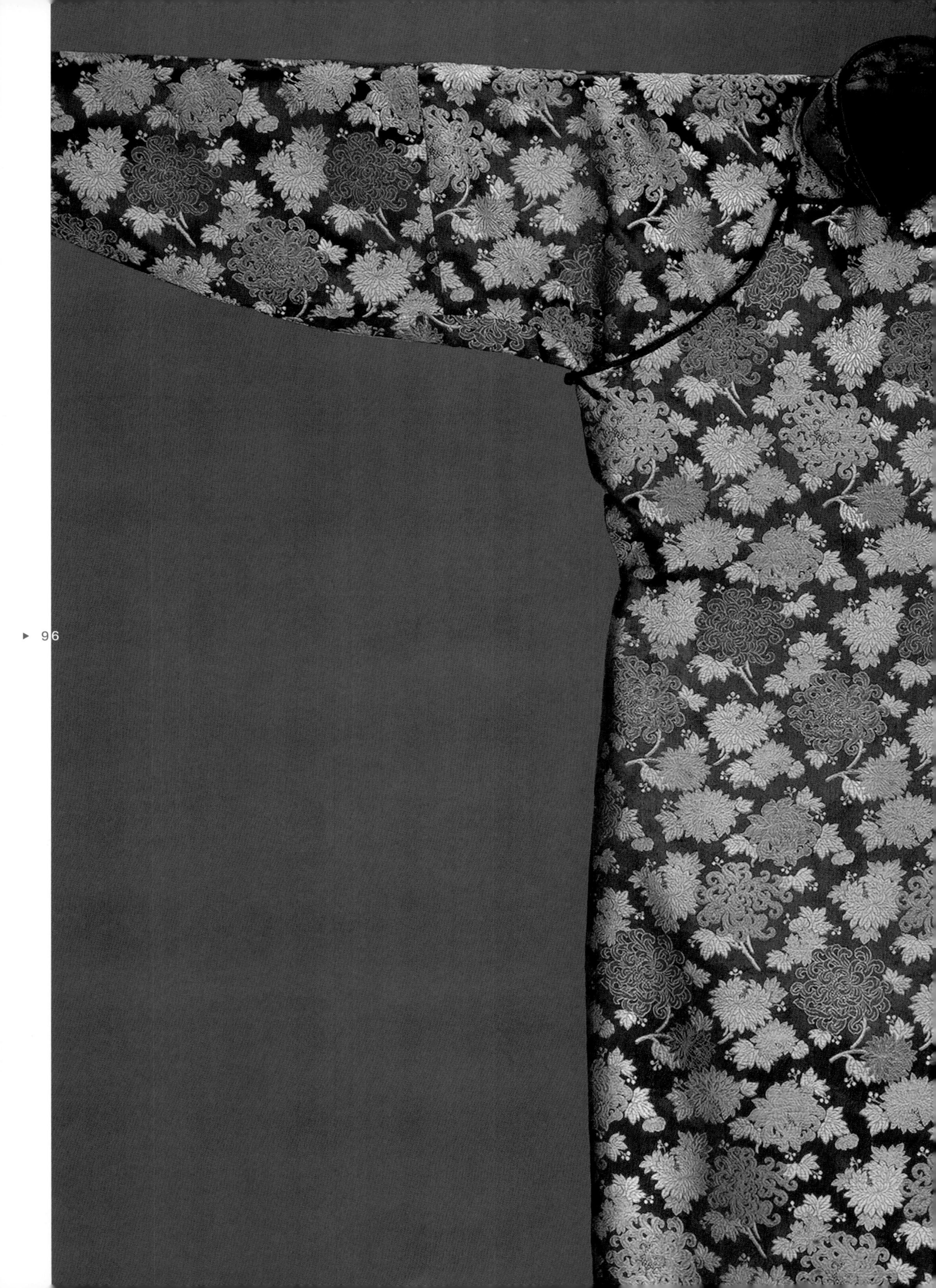

96 ▸ 135 ▸

第二章

现代女性 摩登生活

Chapter Two

Modern Lady Modern Life

第二章 现代女性 摩登生活

新文化运动以后，越来越多的女性走出传统家庭，投身社会活动。她们穿着干练明快而不失优雅气度，打扮入时却也不过分矫饰，旗袍以其简洁的设计和体贴的剪裁，恰到好处地衬托出女性的万种风情，获得她们的青睐。长期以来，女性深受礼教“三绺梳头，两截穿衣”的桎梏，穿着一件制袍服。不用严格地包裹身体，是新女性解放个性的追求。旗袍风尚的兴起，不仅仅是时髦的装扮，更是一种摩登的思想，全新的生活方式。

After the New Culture Movement broke out in 1919, more and more female went out of family and devoted themselves to social activities. They dressed smartly and elegantly, but were not overdressed. Qipao gained its popularity by its simple design and fitting tailoring that perfectly promote to the feminine charm. For a long time, women had been shackled by the traditional ritual. By wearing a robe and loosening the body constraints, the new women sought liberation of their individuality. The trend of qipao was not just related to clothing, but also a form of modern thoughts and lifestyle.

Chapter Two
Modern Lady, Modern Life

一、 台前幕后，几般婀娜

One
On the Screen and Behind the Scene

一、台前幕后，儿般婀娜

上海是中国电影的生产放映基地，电影明星的穿着打扮构建起上海十里洋场灯红酒绿的那一面。喜爱穿着旗袍的女明星很多，她们有的青睐设计独特/出挑惊艳的旗袍，有的偏爱条纹、格纹等西式纹样，颜色素雅。她们穿着旗袍时配以高跟鞋、皮毛大衣，显得婀娜多姿。

胡蝶在银幕上既刻画过逆来顺受的传统女性，也塑造过富有反抗精神的新女性，是名副其实的“电影皇后”。胡蝶出生于上海的一户中产家庭，父亲曾任京奉铁路总稽查。从小随父母多地迁居的经历，养成了她爽利机警的性格。照片中她穿着长过脚面的绣花真丝短袖旗袍，高贵典雅。

One: On the Screen and Behind the Scene

Shanghai was one of the main producing area for Chinese movies. The dressing of movie stars was often seen on the feasting and revelry side of Shanghai. Qipao was favored among female actors: some of them preferred unique and standout design, some preferred plain and simple design with western-style patterns such as stripes or plaids. They would complement qipao with a pair of high-heels and fur coat, heightening the mood of modernity.

Hu Die had portrayed both traditional women who were resigned to the adversity and new women who were rebellious. She was renowned as the “movie queen”. Hu Die was born into a middle class family in Shanghai. Her father was the chief inspector of the Jing Feng railway. Her early experience of moving with her parents since childhood cultivated her bright and alert character. In the picture she was wearing embroidery silk qipao with short sleeves, hemline reaching the floor, noble and elegant.

胡蝶穿着绣花旗袍
Hu Die in embroidery qipao

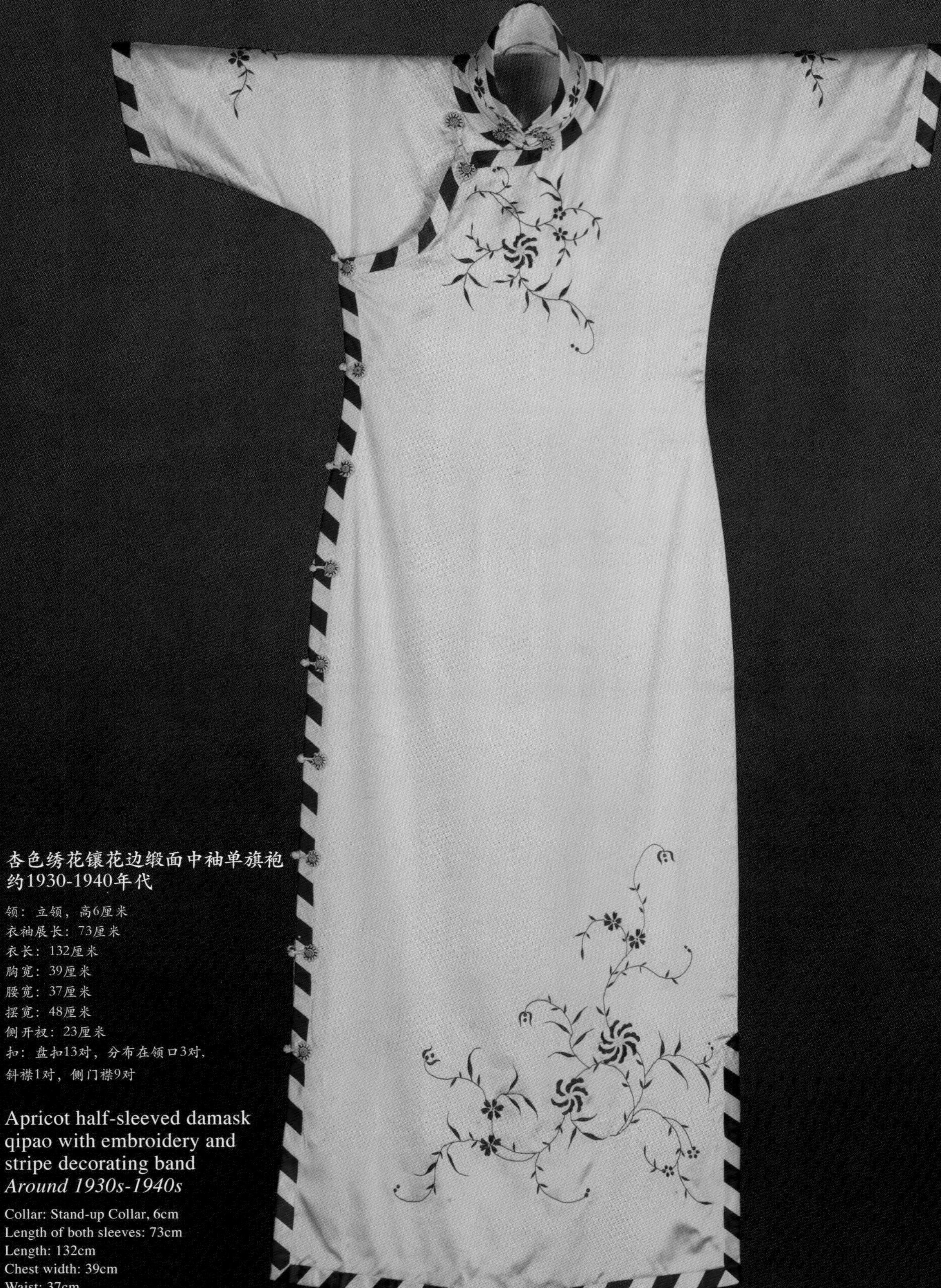

杏色绣花镶花边缎面中袖单旗袍
约1930-1940年代

领：立领，高6厘米
衣袖展长：73厘米
衣长：132厘米
胸宽：39厘米
腰宽：37厘米
摆宽：48厘米
侧开衩：23厘米
扣：盘扣13对，分布在领口3对，斜襟1对，侧门襟9对

Apricot half-sleeved damask qipao with embroidery and stripe decorating band
Around 1930s-1940s

Collar: Stand-up Collar, 6cm
Length of both sleeves: 73cm
Length: 132cm
Chest width: 39cm
Waist: 37cm
Length of hemline: 48cm
Side slits: 23cm
Button: 13 pairs of frog closures, 3 on collar, 1 on slant opening, 9 on the side

咖啡色条纹短袖单旗袍
约1930-1940年代

领：立领，高5.5厘米
衣袖展长：57厘米
衣长：110厘米
胸宽：38厘米
腰宽：35厘米
摆宽：46厘米
侧开衩：18厘米
扣：花式盘扣3对，分布在领口1对，斜襟1对，侧门襟1对；揿纽3对

Coffee stripe qipao with short sleeves
Around 1930s-1940s

Collar: Stand-up Collar, 5.5cm
Length of both sleeves: 57cm
Length: 110cm
Chest width: 38cm
Waist: 35cm
Length of hemline: 46cm
Side slits: 18cm
Button: 3 pairs of floral frog closures, 1 on collar, 1 on slant opening, 1 on the side; 3 metal press studs

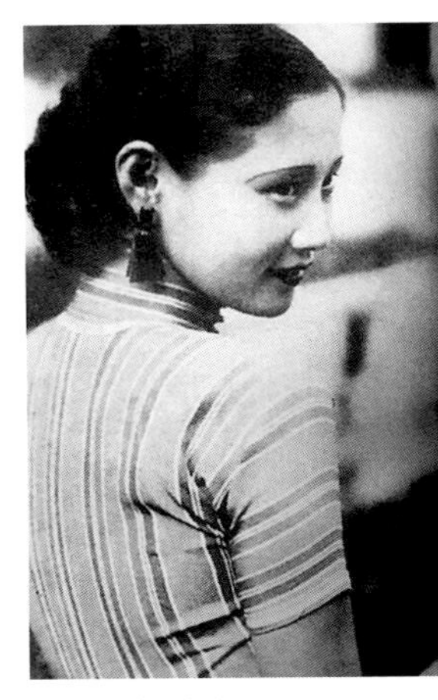

徐来穿着条纹旗袍
Xu Lai in stripe qipao

徐来少时遭遇家变，有一段贫苦难挨的经历。18岁时考入黎锦晖主办的中华歌舞专科学校，毕业后加入明月歌舞团。1933年因主演《残春》而走红。不数年即息影，与国民党将领唐生明结为夫妻，并协助丈夫参加抗日工作。

Xu lai had a hard time living in poverty during her childhood. She entered China Sing and Dancing School at the age of 18, and joined Mingyue (Bright Moon) Troupe. She rose to fame in 1933 for her starring role in the film *Residual Spring*. Within a few years, she retired from the screen and married Tang Sheng-ming, a Kuomintang general, and assisted his husband in anti-Japanese work.

粉色衬绒提花真丝短袖夹旗袍
约1930-1940年代

领：立领，高3.5厘米
衣袖展长：51厘米
衣长：128厘米
胸宽：43厘米
腰宽：41厘米
摆宽：48厘米
侧开衩：35厘米
扣：盘扣3对，分布在领口2对，
斜襟1对；侧门襟10对

Pink silk short-sleeved qipao with velvet lining and woven pattern
Around 1930s-1940s

Collar: Stand-up Collar, 3.5cm
Length of both sleeves: 51cm
Length: 128cm
Chest width: 43cm
Waist: 41cm
Length of hemline: 48cm
Side slits: 35cm
Button: 3 pairs of frog closures, 2 on collar,
1 on slant opening; 9 frog closures,
1 on collar, 1 on the side;
1 metal press stud

黎莉莉穿着同色暗花纹旗袍
Li Lili in shadow pattern qipao

黎莉莉原名钱蓁蓁，是中国共产党地下党员钱壮飞的女儿。黎莉莉是联华影业公司的重要演员，参演《体育皇后》《大路》《渔光曲》等进步影片，塑造了大量青春健美，富有时代气息的新女性，深受观众喜爱。

Li Lili, originally named Qian Zhenzhen. She is the daughter of Qian Zhuangfei who was an underground member of the Communist Party of China. Li Lili was an important actor of Lianhua Film Company, she played in movies as *Queen of Sports*, *The Highway*, and *Song of the Fisherman*. She had created a large number of young and vigorous women and was deeply loved by the audience.

黑地印花真丝无袖单旗袍
约1940年代

领：立领，高5厘米
肩宽：40厘米
衣长：134厘米
胸宽：39厘米
腰宽：38厘米
摆宽：41厘米
侧开衩：37厘米
扣：盘花扣3对，分布在领口1对，斜襟1对，侧门襟1对；揿纽6对

Black ground printed floral pattern silk sleeveless qipao
Around 1940s

Collar: Stand-up Collar, 5cm
Shoulder: 40cm
Length: 134cm
Chest width: 39cm
Waist: 38cm
Length of hemline: 41cm
Side slits: 37cm
Button: 3 pairs of floral frog closures, 1 on collar, 1 on slant opening, 1 on the side; 6 metal press studs

周璇穿着大花无袖旗袍
Zhou Xuan in sleeveless qipao with floral pattern

周璇是老上海著名的歌影明星，莺声燕语般的歌喉打动了无数听众，被誉为“金嗓子”。她主演的《马路天使》是中国电影史上经典佳作，她唱红的《夜上海》从1930年代流行至今。

Zhou Xuan was a famous singer and movie star in 1930s Shanghai. Her singing voice like nightingale had touched numerous audiences and was known as "golden voice". The movie *Angels on the Road* in which she played as leading actress was a classic in Chinese film history, and her famous song *Nightlife in Shanghai* has been popular since the 1930s.

咖啡色烂花绒无袖单旗袍
约1940年代

领：立领，高4厘米
肩宽：37厘米
衣长：110厘米
胸宽：36厘米
腰宽：35厘米
摆宽：42厘米
侧开衩：29厘米
扣：盘花扣3对，分布在领口1对，斜襟1对，侧门襟1对；揿纽5对

Coffee burn-out velvet sleeveless qipao
Around 1940s

Collar: Stand-up Collar, 4cm
Shoulder: 37cm
Length: 110cm
Chest width: 36cm
Waist: 35cm
Length of hemline: 42cm
Side slits: 29cm
Button: 3 pairs of floral frog closures, 1 on collar, 1 on slant opening, 1 on the side; 5 metal press studs

蓝色粗花呢无袖单旗袍
约1940年代

领：立领，高3厘米
肩宽：44厘米
衣长：107厘米
胸宽：42厘米
腰宽：39厘米
摆宽：43厘米
侧开衩：17厘米
扣：盘花扣3对，分布在领口1对、斜襟1对，侧门襟1对；揿纽7对

Blue tweed sleeveless qipao
Around 1940s

Collar: Stand-up Collar, 3cm
Shoulder: 44cm
Length: 107cm
Chest width: 42cm
Waist: 39cm
Length of hemline: 43cm
Side slits: 17cm
Button: 3 pairs of floral frog closures, 1 on collar, 1on slant opening, 1 on the side; 7 metal press studs

阮玲玉是最喜爱旗袍的女明星，不仅生活中以旗袍着装为主，在银幕上塑造的形象，也大都穿着旗袍。通过流传下来的影像资料，可以看到不同的旗袍风格展现着不同人物的身份和性情。1930年4月22日《上海画报》刊登阮玲玉近期照片，卷发，穿着长袖大花纹样旗袍，兼具传统与现代的气息。1934年阮玲玉主演影片《新女性》的宣传照，照片中她穿着纯色提花短袖旗袍，温婉知性的气质与影片主人公韦明极为契合。

Ruan Lingyu is the female star who loved qipao the most. Not only did she wear qipao for daily life, but her figures on movie wore qipao as well. Through the image data passed down, we can see that different styles of qipao show the identities and temperament of different characters. *Pictorial shanghai*, April 22nd 1930, showed a recent photo of Ruan Lingyu. Her hair was in a short wavy coiffure and she was wearing long-sleeved qipao with floral pattern, fusion of tradition and modern. On the promotional photos of Ruan Lingyu's 1934 film *New Women*, she was wearing plain short-sleeved qipao with woven pattern, gentle and soft, looking exactly like her role in the movie.

阮玲玉（载1930年4月22日《上海画报》）
Ruan Lingyu
(Pictorial Shanghai, April 22nd 1930)

《新女性》宣传照
Promotion photo of New Women

黑地衬绒提花镶花边缎面中袖夹旗袍
约1930-1940年代

领：立领，高6厘米
衣袖展长：95厘米
衣长：114厘米
胸宽：39厘米
腰宽：39厘米
摆宽：47厘米
侧开衩：28厘米
扣：盘花扣9对，分布在领口2对，斜襟1对，侧门襟6对；揿纽2对

Black ground half-sleeved damask qipao with velvet lining and bound with lace
Around 1930s-1940s

Collar: Stand-up Collar, 6cm
Length of both sleeves: 95cm
Length: 114cm
Chest width: 39cm
Waist: 39cm
Length of hemline: 47cm
Side slits: 28cm
Button: 9 pairs of floral frog closures, 2 on collar, 1 on slant opening, 6 on the side; 2 metal presse studs

咖啡色衬绒提花真丝中袖夹旗袍
约1930-1940年代

领：立领，高4.5厘米
衣袖展长：67厘米
衣长：129厘米
胸宽：46厘米
腰宽：45厘米
摆宽：51厘米
侧开衩：31厘米
扣：盘花扣2对，分布在领口1对，斜襟1对；盘扣10对，分布在领口1对，侧门襟9对

Coffee silk half-sleeved qipao with velvet lining and woven pattern
Around 1930s-1940s

Collar: Stand-up Collar, 4.5cm
Length of both sleeves: 67cm
Length: 129cm
Chest width: 46cm
Waist: 45cm
Length of hemline: 51cm
Side slits: 31cm
Button: 2 pairs of floral frog closures, 1 on collar, 1 on slant opening; 10 pairs of frog closures, 1 on the collar and 9 on the side

粉色提花真丝短袖单旗袍
约1930-1940年代

领：立领，高6.5厘米
衣袖展长：53厘米
衣长：142厘米
胸宽：36厘米
腰宽：34厘米
摆宽：42厘米
侧开衩：36厘米
扣：盘花扣14对，分布在领口1对，
斜襟1对，侧门襟12对

Pink silk short-sleeved
qipao with woven pattern
Around 1930s-1940s

Collar: Stand-up Collar, 6.5cm
Length of both sleeves: 53cm
Length: 142cm
Chest width: 36cm
Waist: 34cm
Length of hemline: 42cm
Side slits: 36cm
Button: 14 pairs of floral frog closures,
1 on collar, 1 on slant opening,
12 on the side

粉色提花真丝中袖单旗袍
约1930-1940年代

领：立领，高8厘米
衣袖展长：84厘米
衣长：140厘米
胸宽：40厘米
腰宽：40厘米
摆宽：52厘米
侧开衩：30厘米
扣：盘香扣14对，分布在领口3对，斜襟1对，侧门襟10对

Pink half-sleeved silk qipao with woven pattern
Around 1930s-1940s

Collar: Stand-up Collar, 8cm
Length of both sleeves: 84cm
Length: 140cm
Chest width: 40cm
Waist: 40cm
Length of hemline: 52cm
Side slits: 30cm
Button: 14 pairs of incense coil frog closures, 3 on collar, 1 on slant opening, 10 on the side

《野草闲花》剧照
Stage photo of Wild Flowers by The Road

《野草闲花》是著名导演孙瑜在《茶花女》的影响下创作的影片。女主人公丽莲是卖花女出身的歌星，清新素雅的旗袍打扮与她纯真善良、既温顺又勇敢的性格相得益彰。

Wild Flowers by The Road is created by the famous movie director Sun Yu under the influence of *The Lady of the Camellias*. Lillian, the heroine, is a singer born as a flower girl. Her fresh and elegant qipao dress complements her pure and kind-hearted, docile and brave personality.

《三个摩登女性》剧照
Stage Photo of Three Modern Ladies

《三个摩登女性》是田汉编剧，卜万苍导演的影片。阮玲玉主动请缨，请求饰演影片中的进步女性周淑贞。她抹去口红，脱下高跟鞋，从始至终穿着朴素的旗袍，穿梭于工厂和战场，为理想慷慨陈词，把这位具有摩登思想的女工塑造成了经典。

Three Modern Ladies is a film produced by Tian Han, Directed by Pu Wancang. Ruan Lingyu volunteered to play the progressive woman Zhou Shuzhen in the film. She wiped out the lipstick, took off her high heels, and wore a simple qipao from start to finish. She shuttled to the factory and the battlefield, speaking up for her ideals, the modern female worker has been transformed into a classic idol.

灰色印花棉布中袖单旗袍
约1930-1940年代

领：立领，高5厘米
衣袖展长：83厘米
衣长：112厘米
胸宽：49厘米
腰宽：44厘米
摆宽：51厘米
侧开衩：25厘米
扣：盘香扣2对，分布在领口1对；
侧门襟1对；盘花扣1对，在斜襟；揿纽2对

Grey cotton qipao with half sleeves and printed patter
Around 1930s-1940s

Collar: Stand-up Collar, 5cm
Length of both sleeves: 83cm
Length: 112cm
Chest width: 49cm
Waist: 44cm
Length of hemline: 51cm
Side slits: 25cm
Button: 2 pairs of incense coil frog closures, 1 on collar, 1 on slant opening;
2 metal press studs

旗袍的整体风格可能是朴素的，面料可能是廉价的，但细节绝不马虎，盘扣做得一丝不苟，这是海派旗袍的精致与细腻。

The style of qipao may be simple, material could also be economical, but the detail is nothing that could be negligent, and the frog closures are meticulously made, which stands for refinement of Shanghai qipao.

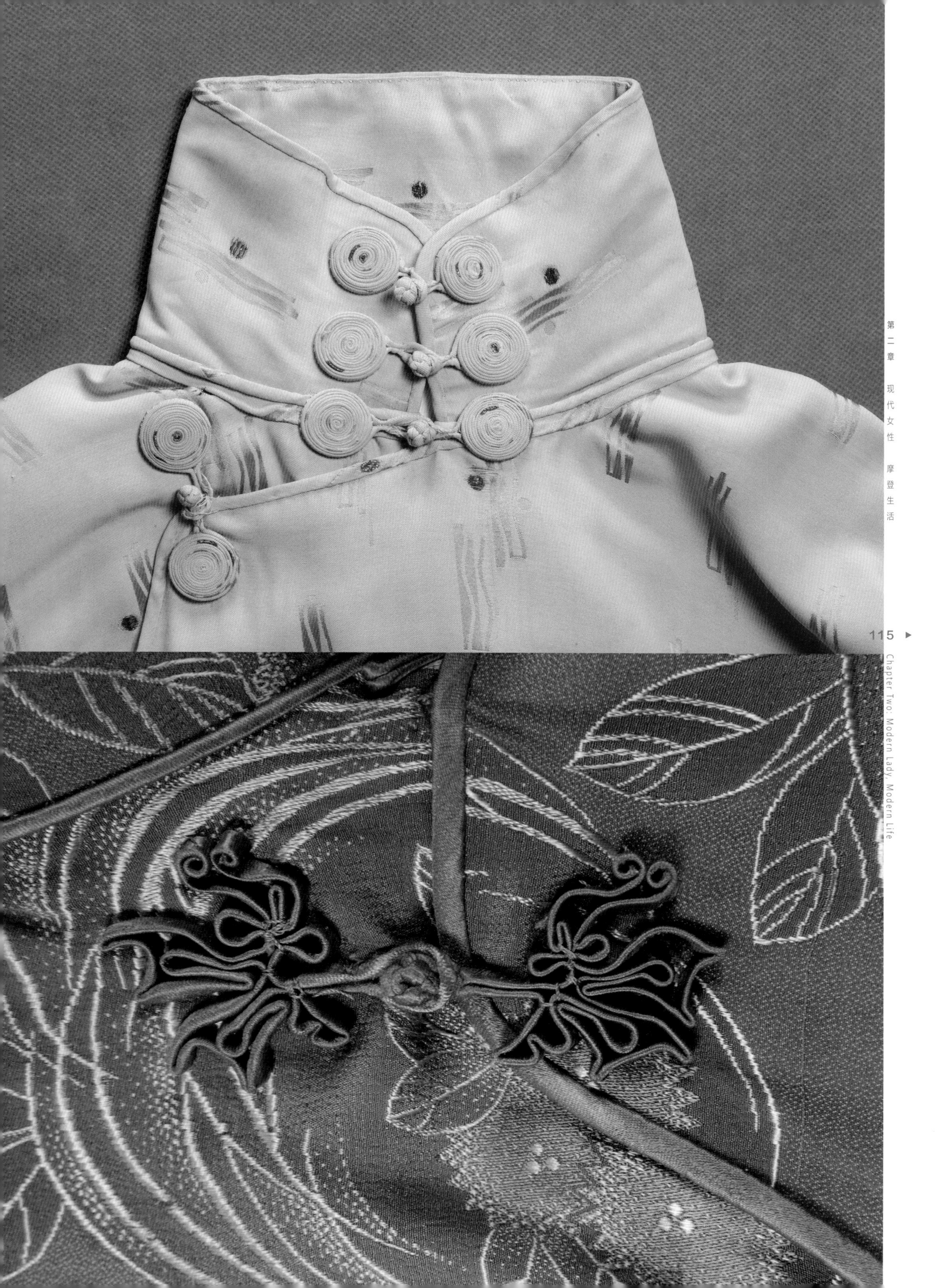

淡粉色衬绒提花真丝短袖夹旗袍
约1930-1940年代

领：立领，高7厘米
衣袖展长：86厘米
衣长：136厘米
胸宽：37厘米
腰宽：36厘米
摆宽：48厘米
侧开衩：31厘米
扣：盘花扣15对，分布在领口3对，
斜襟1对，侧门襟11对

Light pink silk qipao with velvet lining, woven pattern and short sleeves
Around 1930s-1940s

Collar: Stand-up Collar, 7cm
Length of both sleeves: 86cm
Length: 136cm
Chest width: 37cm
Waist: 36m
Length of hemline: 48cm
Side slits: 31cm
Button: 15 pairs of floral frog closures,
3 on collar, 1 on slant opening,
11 on the side

《神女》剧照
Stage photo of Goddess

《神女》是吴永刚编导的第一部作品。他常见没有通告的女演员迫于生计而出卖肉体，充当“神女”。因同情这些女子自食其力却受尽嘲讽与欺辱而创作了这部影片。由阮玲玉、黎铿主演。影片中阮玲玉作为“神女”出场时穿着入时的旗袍，高领窄袖，贴身塑形，长及脚面，开衩过膝，而作为母亲出场时则穿着颜色朴素的旗袍。

Goddess is the first movie directed by Wu Yonggang. He often saw actresses without job sell their bodies for a living and act as “goddess”. The film was made out of sympathy for these women who earned their own living but were suffering from humiliation and bully. Ruan Lingyu and Li Keng were co-starring. In the film, Ruan Lingyu was dressed in trendy qipao with a high collar, narrow sleeves, a figure-hugging fit when she was a “goddess”, a length that reached the ankles and slits exposing the leg from below the knee. While when she was being a mother, she wore qipao in plain color.

咖啡色提花中袖夹旗袍
约1930-1940年代

领：立领，高6.5厘米
衣袖展长：81厘米
衣长：132厘米
胸宽：39厘米
腰宽：37厘米
摆宽：44厘米
侧开衩：30厘米
扣：盘扣12对，分布在领口3对，斜襟1对，侧门襟8对

Coffee half-sleeved qipao with woven pattern and lining
Around 1930s-1940s

Collar: Stand-up Collar, 6.5cm
Length of both sleeves: 81cm
Length: 132cm
Chest width: 39cm
Waist: 37cm
Length of hemline: 44cm
Side slits: 30cm
Button: 12 pairs of frog closures, 3 on collar, 1 on slant opening, 8 on the side

《新女性》剧照
Stage photo of New Woman

《新女性》是著名导演蔡楚生的作品，根据自杀才女艾霞的真实事迹改编。影片中的韦明是一位受过高等教育的知识女性，为了给儿子治病而误入道貌岸然的校董的魔爪。校董凭借尊贵的身份肆意诋毁韦明，她无处投诉，唯有以死证明自己的清白。影片中韦明所穿着的条格纹样旗袍是知识女性的挚爱，既干净素雅，又不失时尚。

The movie *New Woman* was directed by the famous director Cai Chusheng. It was adapted from the story of a talented woman Ai Xia, who committed suicide. Wei Ming in the film was a highly educated woman. She strayed into the hands of a respectable but evil school board in order to cure his son. The school board vilified Wei Ming using his social status; she had nowhere to complain, and could only prove her innocence by death. The check pattern qipao worn by Wei Ming in the film was the favorite of well-educated women. It was simple but trendy.

咖啡地印花真丝中袖单旗袍
约1930-1940年代

领：立领，高7厘米
衣袖展长：76厘米
衣长：130厘米
胸宽：37厘米
腰宽：36厘米
摆宽：43厘米
侧开衩：25厘米
扣：盘花扣13对，分布在领口3对，斜襟1对，侧门襟9对；揿纽1对

Pattern prined silk qipao on coffee ground with half sleeves
Around 1930s-1940s

Collar: Stand-up Collar, 7cm
Length of both sleeves: 76cm
Length: 130cm
Chest width: 37cm
Waist: 36cm
Length of hemline: 43cm
Side slits: 25cm
Button: 13 pairs of floral frog closures, 3 on collar, 1 on slant opening, 9 on the side; 1 metal press stud

二、 对镜写真，描画灵魂

二、对镜写真，描画灵魂

民国时期的上海才女汇集，她们在画布上描摹对生活的感受，画下了许多穿着旗袍的自画像，仿佛是她们灵魂的写真。

方君璧是最早留法的中国女画家，曾在上海举办过轰动一时的画展。蔡元培在为她的画册作序时写道："借欧洲写实之手腕，达中国抽象之气韵。"她所画的《年轻女子》《女子半身像》等，以西洋画的技法，画下了穿着旗袍的中国女性的风韵。

潘玉良对旗袍有很深的情结，画过许多穿着旗袍的自画像，画中的她端庄、娴静，映衬着旗袍的东方风韵，仿佛是她的内心独白。留法艺术家叶星球在《女画家潘玉良》中写道："这位传奇一生、勤苦奋斗的女画家，最后穿着中国的旗袍，遗留下丰富的艺术作品，带着遗憾长眠于法兰西的土地。"

Two: Self-portrait of the Inner-soul

In the period of the Republic of China, the talented women gathered in Shanghai. They depicted the feelings about life on the canvas, and drew many self-portraits with qipao on them, as if it was the portrait of their inner-soul.

Fang Junbi was the first Chinese female painter who studied in France, and she had a sensational exhibition in Shanghai once. Cai Yuanpei wrote the preface for her picture album: 'By means of European realistic techniques, to reach Chinese abstract charm.' Her painting *Young Lady*, *Bust of A Woman* and many more, used western painting techniques and drew the grace of Chinese women in qipao.

Pan Yuliang had a deep affection for qipao, and had painted many self-portraits wearing qipao. In the picture, she is demure and quiet, reflected by the Oriental charm of the qipao, as if it was her inner monologue. Ye Xingqiu, an artist stayed in France wrote in her book *Female Painter Pan Yuliang*: The legendary and hard-working female painter finally wore a Chinese qipao, leaving behind a wealth of works of art and lying in France with regret.

方君璧《年轻女子》
Young Lady by Fang Junbi

方君璧《肖像》
Portrait by Fang Junbi

潘玉良《自画像》（1940年）
Self-portrait by Pan Yuliang (1940)

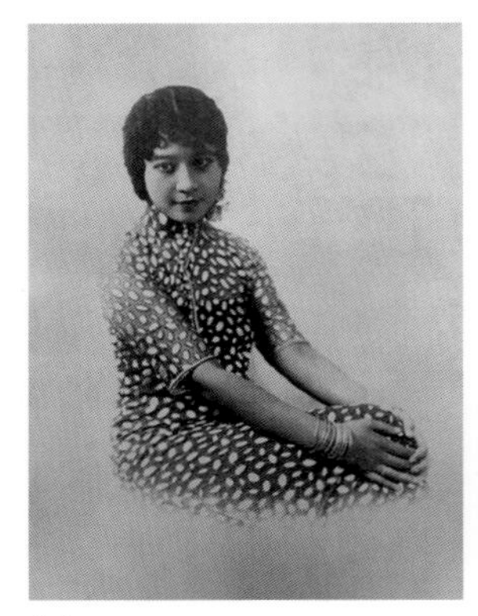

关紫兰
Guan Zilan

关紫兰《少女》
Teenage Girl by Guan Zilan

关紫兰的“野兽派”画风在民国女画家中显得独树一帜。关紫兰生得很美，大熊卓藏《上海西洋画家印象》记录她“悠然而大气，风情万种的大眼睛、略施粉黛的面颊、合身的高领旗袍……戴着西班牙女性中常见的大耳环，摇摇曳曳，每当走动时总会晃动……是中国姑娘中难得一见的美人。”

《L女士像》载于第50期的《良友》上，是她的代表作。少女怀抱洋布狗，红花条纹蓝镶边长袖旗袍，外套蓝色背心，脸上涂着鲜红的胭脂，洋娃娃似的，精致而优雅。

——王璜生、朱青生《女性艺术在中国·自我画像》

Guan Zilan's "fauvism" style of painting is unique among female painters in the Republic of China. Guan Zilan was a real beauty. In the book *West Painting Painters in Shanghai*, it wrote about her as "Relaxed but dignified, flirting big eyes, pinkish cheeks, figure-hugging fit... Wearing large earrings that are common among Spanish women swinging when walking... is a real beauty among Chinese woman."

Portrait of Lady L was published on issue 50th, the Young Companion, and is her representative work. The teenage girl was holding a puppy doll in her arm, wearing long-sleeve qipao with red strip and blue decorative band, covered with blue vest, wearing bright blush, delicate as a doll.

—— Female art in China Self-portrait by Wang Huangsheng and Zhu Qingsheng

孙多慈《孙多慈自画像》（1940）
Self-portrait of Sun Duoci by Sun Duoci

孙多慈《自画像》
Self-portrait by Sun Duoci

孙多慈是徐悲鸿的学生，曾在上海举办画展。她笔下的自己穿着素雅的旗袍，眉宇间流露着女学生般的天真与执着。

Sun Duoci is a student of Xu Beihong. She used to open a painting exhibition in Shanghai. In her drawing, she was wearing plain qipao, with her eyes revealing the innocence and persistence of a student.

黑色烂花绒无袖单旗袍
约1940年代

领：立领，高5.5厘米
肩宽：45厘米
衣长：144厘米
胸宽：38厘米
腰宽：34厘米
摆宽：46厘米
侧开衩：28.5厘米
扣：盘扣12对，分布在领口1对，斜襟1对，侧门襟10对；揿纽1对

Black sleeveless burnt-out velvet qipao
Around 1940s

Collar: Stand-up Collar, 5.5cm
Shoulder: 45cm
Length: 144cm
Chest width: 38cm
Waist: 34cm
Length of hemline: 46cm
Side slits: 28.5cm
Button: 12 pairs of frog closures, 1 on collar, 1 on slant opening, 10 on the side; 1 metal press stud

潘素
Pan Su

张伯驹的夫人潘素是一位擅长工笔的传统画家，时尚的拖地旗袍在她身上演绎的是一番古典气质。董桥写过一篇《永远的潘慧素》，文中描绘潘素穿着旗袍的样子，写道："前几年《老照片》封面上登过潘素一帧三十年代的肖像，亭亭然玉立在一瓶寒梅旁边，长长的黑旗袍和长长的耳坠子衬出温柔的民国风韵：流苏帐暖，春光婉转，冈翁说难得拍得这样传神，几乎听得到她细声说着带点吴音的北京话。"

Zhang Boju's wife, Pan Su, is a traditional painter good at fine brushwork. The fashionable floor-sweeping qipao expressed the classical style on her. Dong Qiao wrote an article about her, *Eternal Pan Huisu*, and described the look of Pan Su in qipao, as: "A few years ago, Pan Su's portrait of the 1930s appeared on the cover of the *Old Photo*. She stood erect and graceful beside a bottle of cold plum, the full-length black qipao and long earrings reflected the tender charm of that time. It was rare to make such a vivid expression. You could almost hear her Beijing dialect with a little accent."

黑色蕾丝短袖单旗袍
约1930-1940年代

领：立领，高5.5厘米
衣袖展长：65厘米
衣长：135厘米
胸宽：40厘米
腰宽：37厘米
摆宽：45厘米
侧开衩：36厘米
扣：盘花扣4对，分布在领口2对，斜襟1对，侧门襟2对

Black short-sleeved qipao in lace fabric
Around 1930s-1940s

Collar: Stand-up Collar, 5.5cm
Length of both sleeves: 65cm
Length: 135cm
Chest width: 40cm
Waist: 37cm
Length of hemline: 45cm
Side slits: 36cm
Button: 4 pairs of floral frog closures, 2 on collar, 1 on slant opening, 2 on the side

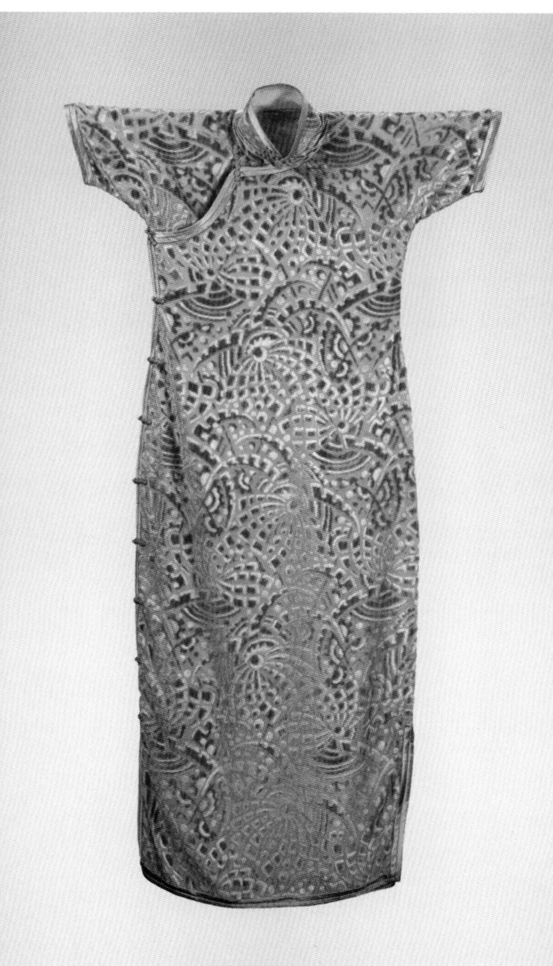

淡咖啡色烂花绒短袖单旗袍
约1930-1940年代

领：立领，高7厘米
衣袖展长：69厘米
衣长：136厘米
胸宽：43厘米
腰宽：42厘米
摆宽：48厘米
侧开衩：27厘米
扣：盘香扣13对，分布在领口3对、
斜襟1对，侧门襟9对

Coffee burnt-out velvet with short sleeves
Around 1930s-1940s

Collar: Stand-up Collar, 7cm
Length of both sleeves: 69cm
Length: 136cm
Chest width: 43cm
Waist: 42cm
Length of hemline: 48cm
Side slits: 27cm
Button: 13 pairs of incense coil frog closures,
3 on collar,
1 on slant opening,
9 on the side

咖啡地提花真丝短袖单旗袍
约1930-1940年代

领：立领，高4厘米
衣袖展长：50厘米
衣长：135厘米
胸宽：43厘米
腰宽：38厘米
摆宽：50厘米
侧开衩：38厘米
扣：盘花扣2对，分布在领口1对，斜襟1对；
盘扣10对，分布在领口1对，侧门襟9对

Coffee burnt-out woven with short sleeves
Around 1930s-1940s

Collar: Stand-up Collar,4cm
Length of both sleeves: 50cm
Length: 135cm
Chest width: 43cm
Waist: 38cm
Length of hemline: 50cm
Side slits: 38cm
Button: 2 pairs of floral frog closures,
1 on collar, 1 on slant opening;
10 pairs of frog closures,
1 on collar, 9 on the side

三、 箧存知性，书此隽永

Three
Elegance Documented in the Book

三、篌存知性，书此隽永

那大概是七月里的一天，张爱玲穿着丝质碎花旗袍，色泽淡雅，也就是当时上海小姐普通的装束。

——柯灵《遥寄张爱玲》

爱玲穿一件桃红的单旗袍，我说好看，她说："桃红的颜色闻得见香气。"

——胡兰成《民国女子》

Three: Elegance Documented in the Book

It was about one day in July, Eileen Chang was wearing a silk Qipao with delicate floral design and elegant color, which was common attire for a Shanghai lady at that time.

—— Miss Eileen Chang by Ke Ling

Eileen Chang was wearing a peach qipao. I said it was nice. She said, "The color of peach could be smelled."

—— The female in the Republic of China by Hu Lancheng

张爱玲手绘《中国人的生活和时装》插图
Chinese life and fashion by Eileen Chang

张爱玲和姑姑
Eileen Chang and her aunt

满地提花无袖单旗袍
约1940年代

领：立领，高3厘米
肩宽：40厘米
衣长：106厘米
胸宽：38厘米
腰宽：37厘米
摆宽：43厘米
侧开衩：15厘米
扣：盘花扣3对，分布在领口1对，斜襟1对，侧门襟1对；揿纽1对

Floral ground sleeveless qipao with woven pattern
Around 1940s

Collar: Stand-up Collar, 3cm
Shoulder: 40cm
Length: 106cm
Chest width: 38cm
Waist: 37cm
Length of hemline: 43cm
Side slits: 15cm
Button: 3 pairs of floral frog closures, 1 on collar, 1 on slant opening, 1 on the side, 1 metal pressed stud

紫地提花真丝无袖单旗袍
约1940年代

领：立领，高2.5厘米
肩宽：44厘米
衣长：110厘米
胸宽：39厘米
腰宽：38厘米
摆宽：48厘米
侧开衩：7厘米
扣：盘扣5对，分布在领口1对，斜襟3对，侧门襟1对；揿纽3对

Woven pattern silk sleeveless qipao in purple ground
Around 1940s

Collar: Stand-up Collar, 2.5cm
Shoulder: 44cm
Length: 110cm
Chest width: 39cm
Waist: 38cm
Length of hemline: 48cm
Side slits: 7cm
Button: 5 pairs of frog closures, 1 on collar, 3 on slant opening, 1 on the side;
3 metal press studs

苏青
Su Qing

总觉着五十年代的上海，哪怕只剩下一个旗袍装，也应当是苏青，因为什么？因为她是张爱玲的朋友。

——王安忆《寻找苏青》

***I** always think that in 1950s Shanghai, even if there is only one person wearing qipao, it should be Su Qing, because what? Because she is a friend of Eileen Chang.*

—— Searching for Su Qing by Wang Anyi

关露穿着旗袍
Guan Lu wearing qipao

她时而着工人装，到工厂区活动，培养骨干；时而浓妆淡抹，着华丽旗袍，出入娱乐场所，传递情报；时而端庄秀丽，如职业妇女，进入日人主办的编辑部。

——丁言昭《"春天里来百花香"——记关露》

***S**he sometimes wearing working suit to the factory, cultivating management candidates; Sometimes she puts on heavy makeup, wearing gorgeous qipao, entering clubs to convey message; Sometime she look professional and goes into the Japanese editorial department.*

—— Flower fragrance in Spring—biography of Guan Lu by Ding Yanzhao

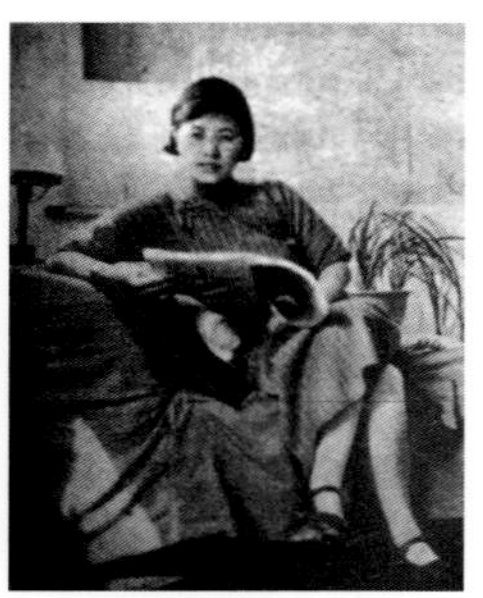

凌叔华穿着旗袍
Ling Shuhua wearing qipao

只见她穿一身淡雅的布旗袍，形单影只的一位中年女教授，被聘到"南大"教华侨子弟。

——浦丽琳《默默的悲凉》

***S**he was dressed in a simple and elegant qipao, a solitary middle-aged female professor, who was hired to teach the children of overseas Chinese in Nanjing University.*

—— Silent Sorrow by Pu Lilin

杨绛穿着旗袍
Yang Jiang wearing qipao

她提起董桥写过《杨绛的旗袍不开衩》，她笑，说，那时是不开衩呀！哪像现在，越开越高，都开到这儿来了！她比画着，我们全都大笑，非常开心。

——陶然《笑眯眯的杨绛》

She mentioned the article Dong Qiao wrote, Yang Jiang's qipao has no slits, and laughed, says, by that time there was really no slits, not like now, the side slits are higher and higher, right up to here! She gestured and we all laughed, how joyful.

—— The smiling Yang Jiang by Tao Ran

张充和穿着旗袍
Zhang Chonghe wearing qipao

她家中衣橱里，挂满风姿妖娆、长短各异的旗袍。

——徐虹《张充和：以慢，以淡》

In her closet at home, there are full of qipao with enchanting charm and various lengths.

—— Zhang Chonghe: Light and Slow by Xu Hong

张允和穿着旗袍，1932年摄于杭州
Zhang Yunhe wearing qipao, taken in Hangzhou

张允和穿着旗袍，1947年摄于美国纽约美术馆
Zhang Yunhe wearing qipao, taken in 1947

恋爱中周有光第一次为我拍照片。我穿着映山红色旗袍，被绿树、绿草拥抱着。

——张允和《曲终人不散》

Zhou Youguang took my first photo since we were in love. I was wearing a red qipao and surrounded by green trees and grass.

—— The Story Ends but People still There by Zhang Yunhe

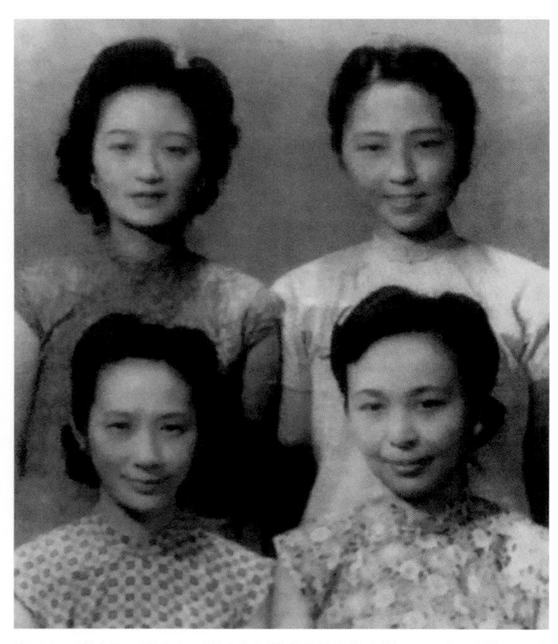

张元和、张允和、张兆和、张充和姐妹穿着旗袍的合影，1946年7月摄于上海
Zhang Yuanhe, Zhang Yunhe, Zhang Zhaohe, Zhang Chonghe group picture wearing qipao. July 1946 in Shanghai

教授们盛赞沈夫人张兆和女士的美丽像阴丹士林一样永不褪色。

——陈友松《无题——怀念沈从文伉俪》诗自注

1933年9月9日，沈二哥三姐在北平中山公园的水榭结婚……姐穿件浅豆沙色普通绸旗袍，沈二哥穿件蓝毛葛的夹袍，是大姐在上海为他们缝制的。

——张允和《三姐夫沈二哥》

The professors praised Shen's wife, Zhang Zhaohe, for her unfading beauty, as indanthrene.

——No title-Miss the couple of Shen Congwen by Chen Yousong from annotation by the author

September 9th, 1933, second brother of Shen and my third sister got married at the waterside of Zhongshan Park, Peiping. My third sister was wearing a brown silk qipao, second brother of Shen wearing a blue lined coat, both ordered from Shanghai by eldest sister.

—— Second Brother of Shen and My Third Sister by Zhang Yunhe

黑地格纹短袖单旗袍
约1940年代

领：立领，高3.5厘米
衣袖展长：51厘米
衣长：112厘米
胸宽：43厘米
腰宽：41厘米
摆宽：46厘米
侧开衩：19厘米
扣：盘香扣9对，分布在领口1对，
斜襟1对，侧门襟7对

Black ground check pattern
qipao with short sleeves
Around 1940s

Collar: Stand-up Collar, 3.5cm
Length of both sleeves: 51cm
Length: 112cm
Chest width: 43cm
Waist: 41cm
Length of hemline: 46cm
Side slits: 19cm
Button: 2 pairs of incense coil frog closures,
1 on collar, 1 on slant opening,
7 on the side

咖啡色提花真丝无袖单旗袍
约1930-1940年代

领：立领，高3厘米
衣袖展长：50厘米
衣长：106厘米
胸宽：40厘米
腰宽：37厘米
摆宽：45厘米
侧开衩：21厘米
扣：盘扣10对，分布在领口1对，
斜襟1对，侧门襟8对

Coffee sleeveless silk qipao
with woven pattern
Around 1930s-1940s

Collar: Stand-up Collar, 3cm
Shoulder: 50cm
Length: 106cm
Chest width: 40cm
Waist: 37cm
Length of hemline: 45cm
Side slits: 21cm
Button: 10 pairs of frog closures,
1 on collar, 1 on slant opening,
8 on the side

咖啡地印花真丝短袖单旗袍
约1930-1940年代

领：立领，高7.5厘米
衣袖展长：58厘米
衣长：132厘米
胸宽：38厘米
腰宽：37厘米
摆宽：47厘米
侧开衩：39厘米
扣：盘花扣10对，分布在领口3对，斜襟1对，侧门襟6对；揿纽6对

Coffee ground pattern printed silk qipao with short sleeves
Around 1930s-1940s

Collar: Stand-up Collar, 7.5cm
Shoulder: 58cm
Length: 132cm
Chest width: 38cm
Waist: 37cm
Length of hemline: 47cm
Side slits: 39cm
Button: 10 pairs of floral frog closures, 3 on collar, 1 on slant opening, 6 on the side; 6 metal press studs

印花真丝短袖单旗袍
约1930-1940年代

领：立领，高3.5厘米
衣袖展长：52厘米
衣长：118厘米
胸宽：43厘米
腰宽：41厘米
摆宽：47厘米
侧开衩：22厘米
扣：盘花扣2对，分布在领口1对，斜襟1对；盘口1对，在侧门襟；揿纽5对

Pattern printed silk short-sleeved qipao
Around 1930s-1940s

Collar: Stand-up Collar,3.5cm
Length of both sleeves: 52cm
Length: 118cm
Chest width: 43cm
Waist: 41cm
Length of hemline: 47cm
Side slits: 22cm
Button: 2 pairs of floral frog closures, 1 on colla, 1 on slant opening; 1 pair of frog closures on the side; 5 metal press studs

136 ▸ 174

第三章

民族工业 上海制造

Chapter Three

National Industry Shanghai Production

第三章　民族工业　上海制造

旗袍的制作是传统的手工艺，旗袍技师身兼设计和制作的职责。精准的量体裁剪，选择适合客人的面料、纹样，加以对流行风尚敏锐的嗅觉，大胆的创新改良，使海派旗袍芳名远播。而近代民族纺织工业在上海蓬勃发展，机织的纺织品和新式印染技术为旗袍的面料增添了一些新的音符，海派旗袍这阕既传统又现代的歌谣余音绕梁，至今不绝。

The production of qipao is a traditional handicraft. Tailors of qipao are responsible for both design and production. Taking accurate body measurements, selecting the right fabric and pattern fit for customers, plus the keen sense of popular fashion, and bold innovation and improvement, all help the reputation of Shanghai style qipao far spread. Meanwhile, the modern national textile industry flourished in Shanghai, with fabric woven by machine and new technology used on dyeing adding new note to the material for making qipao, making the melody of Shanghai style qipao linger and echo till now.

提花印花面料
Woven and printed fabric

一、纺织工业带来的新元素

One
New Elements Brought by Textile Industry

一、纺织工业带来的新元素

传统的海派旗袍多以丝织品作为制作面料。19世纪下半叶，外国商人和民族资本家相继在上海开办缫丝厂，使上海这一刚刚起步的丝织品产地在很短的时间内完成了手工作坊向机器工厂的转变，逐步取代清代江宁织造所在的南京，成为中国丝绸业的中心。进入20世纪后，上海出现了大量丝绸印花厂，丝绸印花的发展大大丰富了丝绸纹样的花色品种。

发达的民族纺织工业使海派旗袍的面料具有多样性和现代性。制作旗袍的面料织、印十分精细，常常是先提花后印花。经典纹样的旗袍显示了上海与欧洲同步的流行节奏，印染工艺又将外来的与本土的因素融合为上海独有的文化符号。

1、阴丹士林和其他印染面料

海派旗袍所用的面料既有传统的纺织品，也有西式的纺织品。传统纺织品多为真丝，花纹则绣花、提花均常见。

One: New Elements Brought by Textile Industry

Traditional Shanghai style qipao was commonly made of silk. In the late 19th century, foreign merchants and ethnic capitalists successively established silk reeling factories in Shanghai, making Shanghai complete the transformation from hand workshops to machine factories, and gradually replaced the place of Nanjing, where Jiangning Weaving institution was located during Qing Dynasty, thus becoming the center of the Chinese silk industry. Since 20th century century, a large number of silk printing factories appeared in Shanghai. The development of silk printing greatly enriched the design and color varieties of silk patterns.

The flourishing national textile industry has made the fabric of Shanghai style qipao diversified and modern. The fabric used to produce qipao was meticulously made, patterns usually first woven then printed. Classical pattern indicates the symmetrical fashion trend of Shanghai with Europe, and the printing and dyeing techniques also integrate foreign and local factors into Shanghai's unique cultural symbols.

1. Indanthrene and Other Dyeing Fabric

Shanghai style qipao uses both traditional and western fabric. Traditional fabric is mostly silk, decorated with embroidery and woven pattern.

提花印花面料
Woven and printed fabric

粉地印花八字襟无袖单旗袍
约1930-1940年代

领：立领，高3.5厘米
肩宽：43厘米
衣长：107厘米
胸宽：40厘米
腰宽：36厘米
摆宽：46厘米
侧开衩：15厘米
扣：揿纽12对

Pink sleeveless qipao with symmetrical opening and printed pattern
Around 1930s-1940s

Collar: Stand-up Collar, 3.5cm
Shoulder: 43cm
Length: 107cm
Chest width: 40cm
Waist: 36cm
Length of hemline: 46cm
Side slits: 15cm
Button: 12 metal press studs

蓝色印花真丝无袖单旗袍
约1940年代

领：立领，高3厘米
肩宽：42厘米
衣长：140厘米
胸宽：40厘米
腰宽：34厘米
摆宽：42厘米
侧开衩：44厘米
扣：盘花扣2对，分布在领口1对，斜襟1对；盘扣1对，在侧门襟；揿纽9对

Blue silk sleeveless qipao with printed pattern
Around 1940s

Collar: Stand-up Collar, 3cm
Shoulder: 42cm
Length: 140cm
Chest width: 40cm
Waist: 34cm
Length of hemline: 42cm
Side slits: 44cm
Button: 2 pairs of floral frog closures, 1 on collar, 1 on slant opening; 1 pair of frog closures on the side; 9 metal press studs

民国晴雨商标阴丹士林布广告画
Advertising poster of Qingyu Brand Indanthrene in Republic of China

阴丹士林布料的广告宣传策略性非常强，几可形成一定程度的广告文化，今存世阴丹士林布制作的旗袍和其他成衣不算多，但流传的阴丹士林广告宣传画非常多，由此可见一斑。

阴丹士林是一种合成染料，1920年代由上海的德国德孚洋行生产，所染衣料经久耐洗，号称“永不褪色”。其中阴丹士林蓝布制作的旗袍在当时深受女学生喜爱，引起一股风潮。

The advertising strategy of Indanthrene cloth is very strong, which can form a certain degree of advertising culture. There are not many qipao or other ready-made clothes made by indanthrene cloth surviving in the present world, but there are many advertisements and posters of indanthrene, which can be evidence.

Indanthrene is a synthetic dyestuff produced in the 1920s by German Defag Company in Shanghai. The dyed material is durable in washing and is said to “never fade”. Among them, the qipao made by indanthrene blue cloth was deeply loved by female students at that time, causing an overwhelming tide.

蓝色阴丹士林布中袖单旗袍
约1930-1940年代

领：立领，高5.5厘米
衣袖展长：89厘米
衣长：124.5厘米
胸宽：41厘米
腰宽：42厘米
摆宽：55厘米
侧开衩：35厘米
扣：盘扣9对，分布在领口2对，
斜襟1对，侧门襟6对

Blue half-sleeved qipao made of indanthrene cloth
Around 1930s-1940s

Collar: Stand-up Collar, 5.5cm
Length of both sleeves: 89cm
Length: 124.5cm
Chest width: 41cm
Waist: 42cm
Length of hemline: 55cm
Side slits: 35cm
Button: 9 pairs of frog closures, 2 on collar, 1 on slant opening, 6 on the side

蓝色阴丹士林布长袖单旗袍
约1930-1940年代

领：立领，高3.5厘米
衣袖展长：131厘米
衣长：110厘米
胸宽：45厘米
腰宽：41厘米
摆宽：54厘米
侧开衩：19厘米
扣：盘扣9对，分布在领口1对，
斜襟1对，侧门襟7对

Blue qipao with long sleeves in indanthrene cloth
Around 1930s-1940s

Collar: Stand-up Collar, 3.5cm
Length of both sleeves: 131cm
Length: 110cm
Chest width: 45cm
Waist: 41cm
Length of hemline: 45cm
Side slits: 19cm
Button: 9 pairs of frog closures,
1 on collar, 1 on slant opening,
7 on the side

蓝色阴丹士林布长袖上袄
约1930-1940年代

领：立领，高6.5厘米
衣袖展长：147厘米
衣长：71厘米
胸宽：49厘米
腰宽：49厘米
摆宽：55厘米
侧开衩：8厘米
扣：盘扣6对，分布在领口1对，
斜襟1对，侧门襟4对

Blue long-sleeved lined-jacket
in indanthrene cloth
Around 1930s-1940s

Collar: Stand-up Collar, 6.5cm
Length of both sleeves: 147cm
Length: 71cm
Chest width: 49cm
Waist: 49cm
Length of hemline: 55cm
Side slits: 8cm
Button: 6 pairs of frog closures,
1 on collar, 1 on slant opening,
4 on the side

2、几何纹样和其他西式审美的影响

常见于旗袍面料的传统纹样有牡丹纹、兰花纹、万字纹、龙凤纹等。受西式审美影响而产生的纹样则有玫瑰纹、火腿纹、扇子纹、羽毛纹等。此外，条纹、格纹也是极为时尚的西式纹样。1936年第257期《玲珑》描绘最新旗袍式样采用的西式纹样，是"精细底而非粗枝大叶底，颜色一律是白底而加上淡红的、淡黄的、淡蓝的花纹"。

20世纪初期的上海受到迪考艺术影响很深，这样的风格不仅常见于建筑、家具的设计，也融入旗袍的制作中，是最具海派特色的艺术风格之一。

2. Geometric Pattern and Other Design Influenced by Western Aesthetics

The patterns commonly seen in qipao fabrics include traditional pattens such peony, orchid, swastikas, dragon and phoenix. Patterns influenced by western aesthetic are roses, paisley, fans and feathers. In addition, stripes and plaid are also very fashionable Western patterns. Described in *The Linloon Magazine*, issue 257th, 1936, the latest qipao using western pattern is "exquisite white ground with light red, yellow and blue pattern".

Shanghai in the early 20th century was deeply influenced by Art Deco, the style was not only found in designs of construction, furniture, but also merged in the production of qipao, which then formed the artistic style with the most Shanghai style character.

龙凤纹样
Dragon and phoenix pattern

黑地提花真丝长袖夹旗袍
约1940年代

领：立领，高6厘米
衣袖展长：122厘米
衣长：119厘米
胸宽：42厘米
腰宽：37厘米
摆宽：44厘米
侧开衩：23厘米
扣：盘扣3对，分布在领口1对，斜襟1对，侧门襟1对；揿纽7对

Black ground long-sleeved silk qipao with lining and woven pattern
Around 1940s

Collar: Stand-up Collar, 6cm
Length of both sleeves: 122cm
Length: 119cm
Chest width: 42cm
Waist: 37cm
Length of hemline: 44cm
Side slits: 23cm
Button: 3 pairs of frog closures, 1 on collar, 1 on slant opening, 1 on the collar;
7 metal press studs

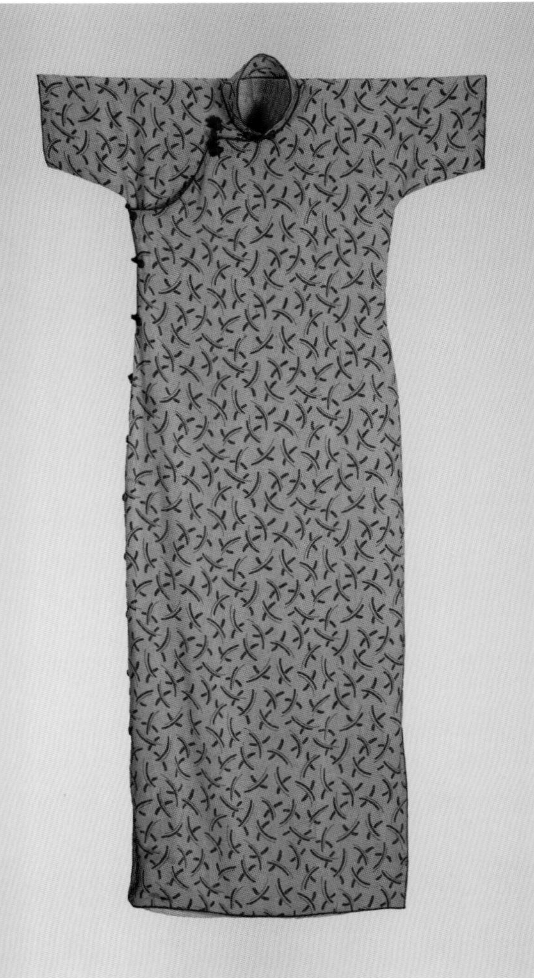

咖啡色条纹短袖单旗袍
约1940年代

领：立领，高5.5厘米
衣袖展长：66厘米
衣长：128厘米
胸宽：40厘米
腰宽：40厘米
摆宽：47厘米
侧开衩：39厘米
扣：盘香扣10对，分布在领口2对，
斜襟1对，侧门襟7对

Coffee short-sleeved qipao
with herringbone pattern
Around 1940s

Collar: Stand-up Collar, 5.5cm
ength of both sleeves: 66cm
Length: 128cm
Chest width: 40cm
Waist: 40cm
Length of hemline: 47cm
Side slits: 39cm
Button: 10 pairs of incense coil frog closures,
2 on collar, 2 on slant opening,
7 on the side

咖啡色印花中袖单旗袍
约1930-1940年代

领：立领，高5.5厘米
衣袖展长：69厘米
衣长：133.5厘米
胸宽：40厘米
腰宽：38厘米
摆宽：47厘米
侧开衩：28.5厘米
扣：盘香扣12对，分布在领口2对，
侧门襟10对；盘花扣1对，在斜襟

Coffee half-sleeved qipao
with printed pattern
Around 1930s-1940s

Collar: Stand-up Collar, 5.5cm
Length of both sleeves: 69cm
Length: 133.5cm
Chest width: 40cm
Waist: 38cm
Length of hemline: 47cm
Side slits: 28.5cm
Button: 12 pairs of incense coil frog closures,
2 on collar, 10 on slant opening;
1 pair of floral frog closures on the side

紫地火腿纹印花真丝无袖旗袍
约1940年代

领：立领，高4.5厘米
肩宽：44厘米
衣长：134厘米
胸宽：42厘米
腰宽：39厘米
摆宽：43厘米
侧开衩：28厘米
扣：盘扣3对，分布在领口1对，
门襟1对，侧门襟1对；揿纽7对

Silk sleeveless qipao
with printed paisley pattern
on purple ground
Around 1940s

Collar: Stand-up Collar, 4.5cm
Shoulder: 44cm
Length: 134cm
Chest width: 42cm
Waist: 39cm
Length of hemline: 43cm
Side slits: 28cm
Button: 3 pairs of frog closures,
1 on collar, 1 on slant opening, 1 on the side;
7 metal press studs

火腿纹是佩兹利纹样在中国常见的名称。佩兹利图案因工业革命时期英国小镇佩兹利得名。关于它的起源较众说纷纭，目前为通行一种，是起源于印度克什米尔地区生产的披肩图案。中国古代新疆地区的织物上就已经有佩兹利图案，常称为巴坦姆图案。民国时期，佩兹利图案在内地生产的丝织品上出现并流行，有单独构图，有组合、变形，图案的造型已相当丰富。

Ham pattern is a commonly used local name for paisley pattern in China. Paisley pattern was named after a British town during the industrial revolution. The origin of it remains uncertain, but the most popular explanation today, is that it was originated from a shawl pattern produced in Kashmir, India. Same pattern was also found in ancient China on fabrics around Xinjiang region. During the period of the Republic of China, paisley pattern appeared and became trendy on silk fabric in the mainland. The pattern could be used as its own, or as combination or variation, the designs were abundant and various.

咖啡地火腿纹印花真丝无袖单旗袍
约1940年代

领：立领，高4.5厘米
衣袖展长：47厘米
衣长：123厘米
胸宽：44厘米
腰宽：42厘米
摆宽：44厘米
侧开衩：28厘米
扣：盘花扣3对，分布在领口1对，
斜襟1对，侧门襟1对

Silk sleeveless qipao with printed paisley pattern on coffee ground
Around 1940s

Collar: Stand-up Collar, 4.5cm
Shoulder: 47cm
Length: 123cm
Chest width: 44cm
Waist: 42cm
Length of hemline: 44cm
Side slits: 28cm
Button: 3 pairs of floral frog closures,
1 on collar, 1 on slant opening, 1 on the side

蓝地火腿纹印花真丝长袖夹旗袍
约1940年代

领：立领，高5.5厘米
衣袖展长：132厘米
衣长：120厘米
胸宽：46厘米
腰宽：43厘米
摆宽：47厘米
侧开衩：28厘米
扣：领口钩子1对，揿纽5对，
拉链1条

Blue silk long-sleeved lined qipao with printed paisley pattern
Around 1940s

Collar: Stand-up Collar, 5.5cm
Length of both sleeves: 132cm
Length: 120cm
Chest width: 46cm
Waist: 43cm
Length of hemline: 47cm
Side slits: 28cm
Button: 1 hook-and-eye clasp on the collar, 5 metal press studs, 1 zip

绿地印花真丝短袖单旗袍
约1940年代

领：立领，高6.5厘米
衣袖展长：68厘米
衣长：121厘米
胸宽：45厘米
腰宽：39厘米
摆宽：40厘米
侧开衩：38厘米
扣：揿纽5对

Silk short-sleeved qipao with printed pattern on green ground
Around 1940s

Collar: Stand-up Collar, 6.5cm
Length of both sleeves: 68cm
Length: 121cm
Chest width: 45cm
Waist: 39cm
Length of hemline: 40cm
Side slits: 38cm
Button: 5 metal press studs

二、非遗技艺，匠人精神

Two
Intangible Skills, Craftsman Spirit

二、 非遗技艺，匠人精神

除了“镶、嵌、滚、宕、盘、绣”的传统服饰制作技法之外，旗袍的制作十分注重量体裁衣，关注穿着者的气质和需求。2011年5月，以海派旗袍为代表的中式服装制作技艺入选《第三批国家级非物质文化遗产名录》，昔日的时尚在巧夺天工的匠人手中成为今天的传统和经典。

1、从平面到立体，从归拔到省道

海派旗袍经历了从1920年代平面裁剪到1940年代立体裁剪的转变，这种转变伴随着女性胸部的逐渐解放而产生，变化的脚步温和而曲折。1928年7月15日第76期《常识》报刊文提到：“今日上海之女性，已十九解放矣，往来于道者，大都双峰高耸……男女双方均已司空见惯，实无羞耻之理。”女性的身体特征受到尊重，自然的体态获得欣赏，旗袍的制作则开始采用了归拔、省道的技法，从适应人体的曲线到着力刻画表现曲线的美感。

1930年1月6日第329期《江苏省政府公报》以“既伤肢体，复害卫生，弱种弱国，贻害无穷”为由颁布禁令，禁止妇女“缠足、束腰、穿耳、束胸”。

Two: Intangible Skills, Craftsman Spirit

Except the traditional skills of binding, insertion, bounding, bias binding, braiding and embroidery, qipao also pay great attention to customization, focusing on the temperament and needs of wearers. In May 2011, the Chinese garment making technique, represented by Shanghai style qipao, was selected on *The Third Batch of National Intangible Cultural Heritage Lis*. The fashion from past has become the tradition and the classic in the hands of the craftsman.

1. From Two-dimension to Three-dimension, From Ironing and Stretching to Darts

The tailoring for Shanghai style qipao has experienced the change from two-dimensional in 1920 to three-dimension in 1940. This transformation was accompanied by the gradual liberation of women's breasts, and the steps of change were gentle and tortuous. *Common sense*, issue *76* which was published on July 15th, 1928 wrote: “The majority of Shanghai female today are liberated now, the ones we see on the street are all showing their natural curve. There is no reason to be shame as both men and women are so accustomed.” Women's physical characteristics were respected and their natural bodies were appreciated, the production of qipao began to integrate into techniques such as ironing and stretching technique and dart, from adapting to the curve of human body to boldly expressing the beauty of curve.

Published on January 6th, 1930, *Gazette of Jiangsu Provincial Government* issued an injunction based on the reason of “hurting body and harming for hygiene, weakening people and the country”, prohibiting women from “foot binding, waist girdle, ear piercing and chest corset”.

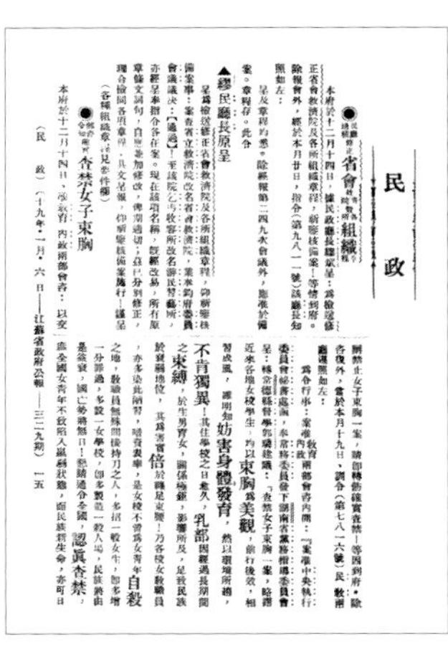

民　政

●民廳呈請修正省會救濟院及各所組織章程令

本府於十二月十四日，據民政廳長繆斌呈：為檢送修正省會救濟院及各所組織章程，祈鑒核備案！等情到府。除報會外，經於本月廿日，指令（第九八一一號）該廳長知照如左：

呈及章程均悉。除經報第二四九次會議外，應准於備案。章程存。此令

▲繆民廳長原呈

呈為檢送修正省會救濟院及各所組織章程，仰祈鑒核備案事：案查省立救濟院改名省會救濟院，業奉鈞府委員會議議決：【通過】！至該院乙丙教養所改名游民習藝所，亦經呈奉指令各在案。現在該項名稱，既經改易，所有原章條文詞句，自應兼加修改，俾期適切；茲已分別修正，理合檢同各項章程，具文呈報，仰祈鑒核備案施行！謹呈

（各種組織章程見參件欄）

●教育部令知內政部會咨查禁女子束胸

本府於十二月十四日，准教育 內政兩部會咨：以安嚴禁止女子束胸一案，請即轉飭確實查禁！等因到府。除咨復外，當於本月十九日，訓令（第七八一六號）民 教兩廳遵照如左：

為令行事：案准教育 內政兩部會咨內開：「案准中央執行委員會秘書處函，奉常務委員發下湖南省黨務指導委員會呈：轉據常德縣督學郭璟建議：『查禁女子束胸一案，略謂近來各地女校學生，均以束胸為美觀，前行後效，相習成風，詎明知妨害身體發育，然以環境所趨，不肯獨異！其住學校之日愈久，乳部因經過長期間之束縛，於生男育女，關係綦鉅，影響所及，足致民族於衰弱地位，其為害實倍於纏足束腰！乃各校女教職員，亦多染此陋習，[illegible]表率，是女校不啻為女青年自殺之地，教職員無殊間接持刀之人，多招一般女生，即多增一分罪過，多設一女學校，即多製造一殺人場，民族將由是衰弱，國亡勢將無日！懇請通令全國，認真查禁，庶全國女青年不致陷入羸弱狀態，而民族將生命，亦可日

（民　政）（十九年•一月•六日——江蘇省政府公報——三二九期）　一五

《查禁女子束胸令》，载1930年1月6日第329期《江苏省政府公报》
Injunction for women corset, published on issue 329th, January 6th, 1930, Gazette of Jiangsu Provincial Government

1920年代的女性多仍奉行束胸的方法，将胸部压得平板。旗袍继承传统服饰的直线造型，大致符合女性的身形线条。1930年代随着女性束胸痼习的逐步消除，在旗袍的制作中出现了归拔的技法，即在裁剪后归拢或拔开织物进行定型熨烫，通过热塑变形使旗袍更贴合人体曲线。用这种技法制作的旗袍曲线较为柔和，符合女性胸部逐渐放开的需求。1940年代旗袍的制作开始运用西式裁剪“开省道”的方式，有了胸省或腰省的旗袍曲线更为分明。

Around the 1920s, most women were still using chest corset, to make the breast flat. Qipao inherits the linear shape of traditional costume and roughly conforms to the body line of women. In the 1930s, with the gradual elimination of the female breast binding, the ironing and stretching technique was applied during the production of qipao, which is basically to pull in or push out a contour when ironing to fit the dress to the contour of the wearer's figure. The qipao made with this technique has a softer curve, which meets the needs of the gradual opening of women's breasts. During 1940s, the western tailoring feature of darts was introduced to qipao making, the silhouette of qipao was more curvy and feminine with bust darts and waist darts.

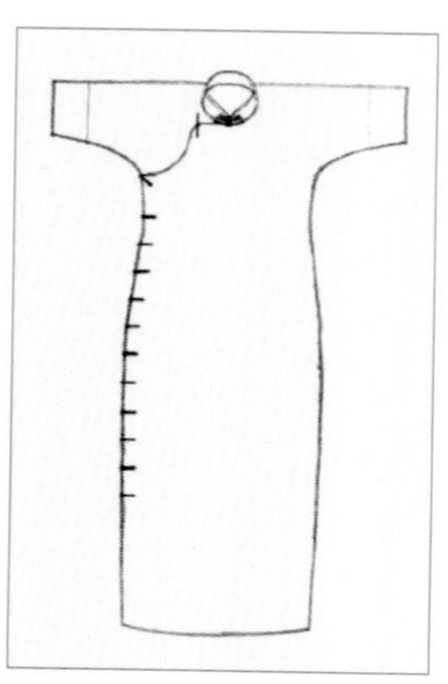

旗袍造型结构图示一：
直线裁剪

Structure of qipao figure 1: linear cutting

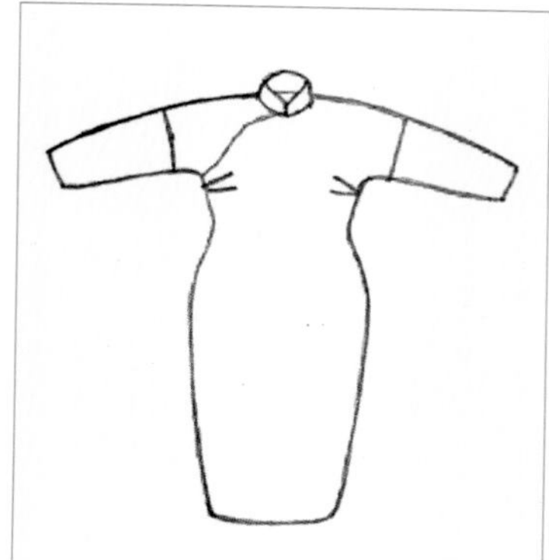

旗袍造型结构图示二：
曲线裁剪，归拔技法

Structure of qipao figure 2: curvy cutting, using ironing and stretching technique

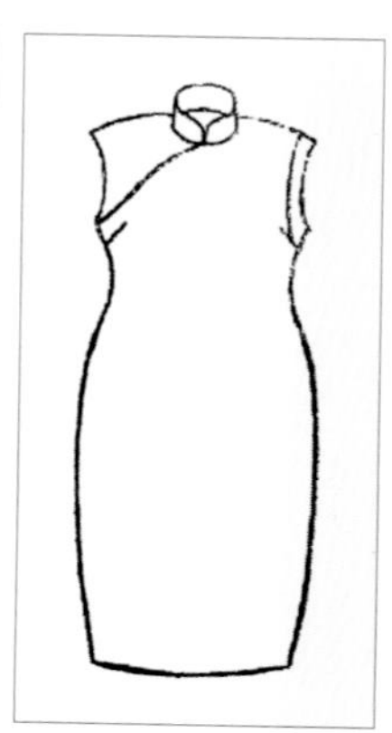

旗袍造型结构图示三：
立体裁剪，胸省腰省

Structure of qipao figure 3: three-dimensional cutting, using bust darts and waist darts

咖啡地印花真丝中袖单旗袍
约1930-1940年代

领：立领，高5.5厘米
衣袖展长：66厘米
衣长：128厘米
胸宽：40厘米
腰宽：40厘米
摆宽：47厘米
侧开衩：39厘米
扣：盘扣10对，分布在领口2对，斜襟2对，侧门襟6对

Coffee ground printed half-sleeved silk qipao
Around 1930s-1940s

Collar: Stand-up Collar, 5.5cm
Length of both sleeves: 66cm
Length: 128cm
Chest width: 40cm
Waist: 40cm
Length of hemline: 47cm
Side slits: 39cm
Button: 10 pairs of frog closures, 2 on collar, 2 on slant opening, 6 on the side

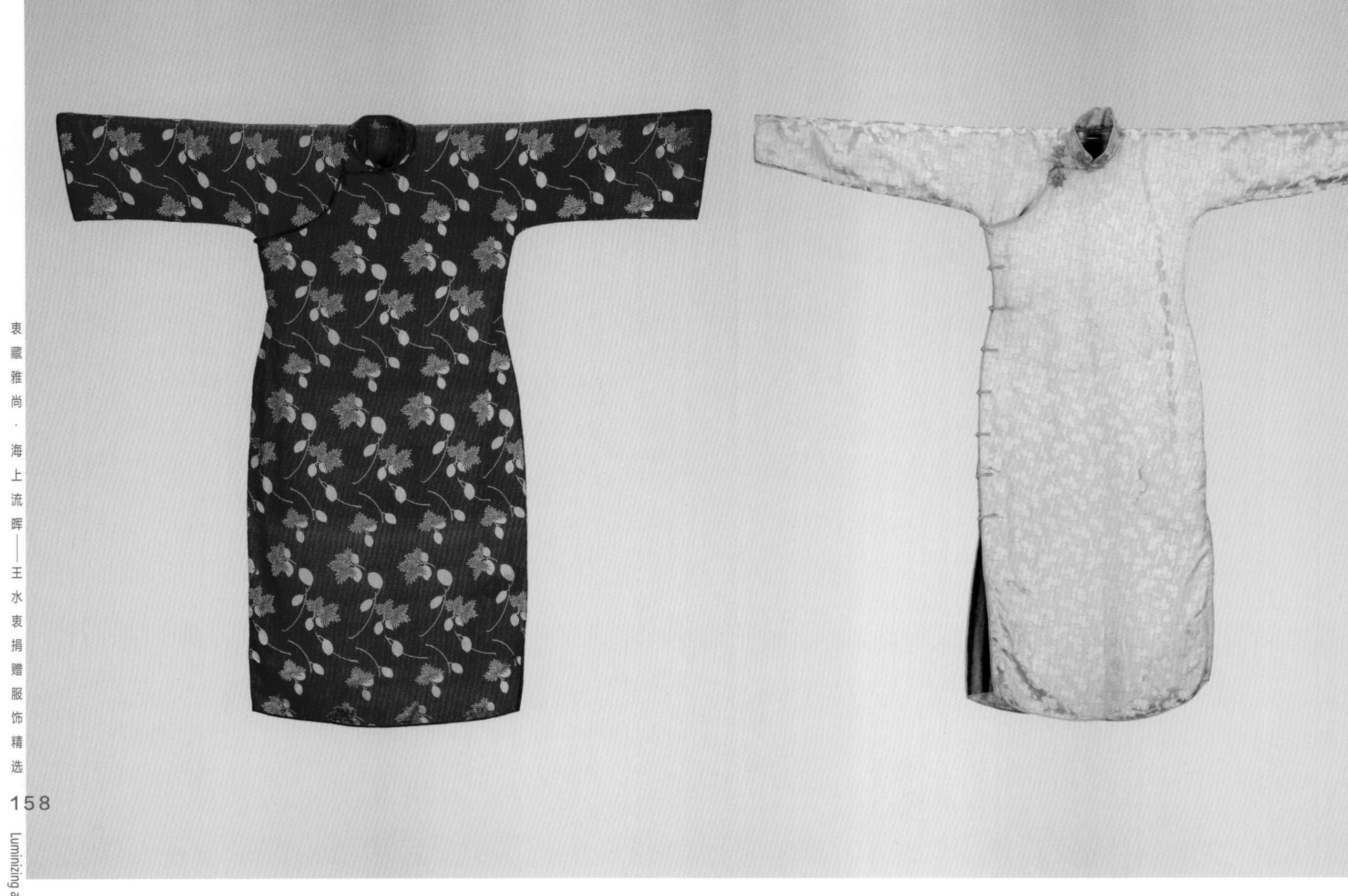

紫色衬绒提花真丝长袖夹旗袍
约1930-1940年代

领：立领，高2厘米
衣袖展长：109厘米
衣长：109厘米
胸宽：43厘米
腰宽：41厘米
摆宽：46厘米
侧开衩：12厘米
扣：盘扣3对，分布在领口1对，斜襟1对，侧门襟1对；揿纽6对

Purple silk long-sleeved qipao with woven pattern and velvet lining
Around 1930s-1940s

Collar: Stand-up Collar, 2cm
Length of both sleeves: 109cm
Length: 109cm
Chest width: 43cm
Waist: 41cm
Length of hemline: 46cm
Side slits: 12cm
Button: 3 pairs of frog closures, 1 on collar, 1 on slant opening, 1 on the side; 6 metal press studs

杏色提花缎长袖夹旗袍
约1940年代

领：立领，高4.5厘米
衣袖展长：122厘米
衣长：114厘米
胸宽：39厘米
腰宽：36厘米
摆宽：45厘米
侧开衩：35厘米
扣：盘花扣2对，分布在领口1对，斜襟1对；盘扣9对，分布在侧门襟8对；揿纽6对

Apricot damask long-sleeved lined qipao with woven pattern
Around 1940s

Collar: Stand-up Collar, 4.5cm
Length of both sleeves: 122cm
Length: 114cm
Chest width: 39cm
Waist: 36cm
Length of hemline: 45cm
Side slits: 35cm
Button: 2 pairs of floral frog closures, 1 on collar, 1 on slant opening;
9 pairs of frog closures, on the side; 6 metal press studs

黑色提花印花长袖单旗袍
约1940年代

领：立领，高6厘米
衣袖展长：140厘米
衣长：131厘米
胸宽：45厘米
腰宽：40厘米
摆宽：41厘米
侧开衩：43厘米
扣：盘香扣3对，分布在领口1对、斜襟1对、侧门襟1对；揿纽2对

Black long-sleeved qipao with woven and printed pattern
Around 1940s

Collar: Stand-up Collar, 6cm
Length of both sleeves: 140cm
Length: 131cm
Chest width: 45cm
Waist: 40cm
Length of hemline: 41cm
Side slits: 43cm
Button: 3 pairs of frog closures, 1 on collar, 1 on slant opening, 1 on the side;
2 metal press studs

满地提花真丝无袖单旗袍
约1940年代

领：立领，高4厘米
肩宽：43厘米
衣长：107厘米
胸宽：44厘米
腰宽：39厘米
摆宽：46厘米
侧开衩：10厘米
扣：盘扣3对，分布在领口1对、斜襟1对，侧门襟1对；揿纽7对

Silk sleeveless qipao with full woven pattern
Around 1940s

Collar: Stand-up Collar, 4cm
Shoulder: 43cm
Length: 107cm
Chest width: 44cm
Waist: 39cm
Length of hemline: 46cm
Side slits: 10cm
Button: 3 pairs of frog closures, 1 on collar, 1 on slant opening, 1 on the side;
7 metal press studs

2、量体裁衣，度身定制

海派旗袍尤重量体裁衣，制作一件旗袍需要量取身上二十多个部位的尺寸。旗袍师傅不仅关注客人的身体特征，还了解客人的气质、偏好，熟知客人经常出席的场合。细腻的心思、娴熟的技艺、平和的心态将匠人精神发挥到极致。

已故旗袍大师褚宏生是海派旗袍制的代表人物，老人一直清晰地记得老顾客的喜好。比如，胡蝶喜欢镶边的旗袍，有时兴起，还会自己动手设计。褚老曾为她制作一件翠绿色蝴蝶图案的软缎旗袍。白光喜爱深色缎面旗袍，袖子做得很短，露出白皙的双臂。陈香梅喜欢收缩性好，手感柔软的真丝面料。王光美喜欢古典花色的旗袍，暗色大花或是纯色素色都是她的心头好。

2. Body Measurement and Customize

Making a Shanghai style qipao requires very detailed body measurements, generally more than 20 measurements have to be taken for making one qipao. The tailor has to not only note the body figure, but also the character and hobby of the customer, as well as knowing their frequent venue. Thoughtful ideas, skillful technique and peaceful mind made the zenith of craftsmanship.

Chu Hongsheng, the late qipao master, was the representative figure of the Shanghai style qipao tailor. He remembered clearly about the preference of frequent customers. For example, famous actress Hu Die preferred qipao with decorative edging, she would even design her own qipao. Master Chu used to make her an emerald green satin qipao with butterfly motif. Bai Guang preferred dark damask qipao, with very short sleeves to expose her radiant white arms. Chen Xiangmei preferred stretchable soft silk fabric. Wang Guangmei liked qipao with classic pattern, dark floral and plain were her favorite.

淡黄色提花无袖真丝单旗袍
约1940年代

领：立领，高3厘米
肩宽：46厘米
衣长：121厘米
胸宽：42厘米
腰宽：41厘米
摆宽：48厘米
侧开衩：17厘米
扣：盘花扣3对，分布在领口1对，斜襟1对，侧门襟1对；揿纽6对

Pale yellow sleeveless silk qipao with woven pattern
Around 1940s

Collar: Stand-up Collar, 3cm
Shoulder: 46cm
Length: 121cm
Chest width: 42cm
Waist: 41cm
Length of hemline: 48cm
Side slits: 17cm
Button: 3 pairs of floral frog closures, 1 on collar, 1 on slant opening;
6 metal presss studs

绿色提花真丝中袖单旗袍
约1930-1940年代

领：立领，高8厘米
衣袖展长：69厘米
衣长：131厘米
胸宽：38厘米
腰宽：37厘米
摆宽：47厘米
侧开衩：27厘米
扣：盘扣16对，分布在领口4对，斜襟2对，侧门襟10对

Green half-sleeved silk qipao with woven pattern
Around 1930s-1940s

Collar: Stand-up Collar, 8cm
Length of both sleeves: 69cm
Length: 131cm
Chest width: 38cm
Waist: 37cm
Length of hemline: 47cm
Side slits: 27cm
Button: 16 pairs of frog closures, 4 on collar, 2 on slant opening, 10 on the side

黑地提花真丝短袖单旗袍
约1930-1940年代

领：立领，高3.5厘米
衣袖展长：52厘米
衣长：104厘米
胸宽：43厘米
腰宽：42厘米
摆宽：48厘米
侧开衩：17厘米
扣：盘扣10对，分布在领口1对，斜襟1对，侧门襟8对

Black ground short-sleeved silk qipao with woven pattern
Around 1930s-1940s

Collar: Stand-up Collar, 3.5cm
Length of both sleeves: 52cm
Length: 104cm
Chest width: 43cm
Waist: 42cm
Length of hemline: 48cm
Side slits: 17cm
Button: 10 pairs of frog closures, 1 on collar, 1 on slant opening, 8 on the side

墨绿色提花真丝中袖单旗袍
约1930-1940年代

领：立领，高4厘米
衣袖展长：56厘米
衣长：129厘米
胸宽：40厘米
腰宽：37厘米
摆宽：46厘米
侧开衩：39厘米
扣：盘花扣9对，分布在领口1对，
斜襟1对，侧门襟7对；揿纽2对

Dark green half-sleeved silk qipao with woven pattern
Around 1930s-1940s

Collar: Stand-up Collar, 4cm
Length of both sleeves: 56cm
Length: 129cm
Chest width: 40cm
Waist: 37cm
Length of hemline: 46cm
Side slits: 39cm
Button: 9 pairs of floral frog closures,
1 on collar, 1 on slant opening, 7 on the side;
2 metal press studs

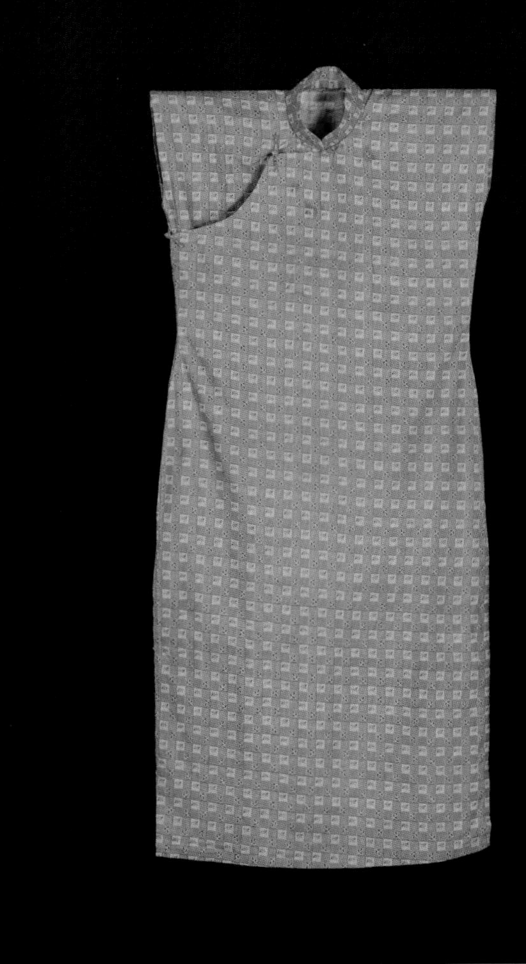

咖啡色提花真丝中袖夹旗袍
约1930-1940年代

领：立领，高6厘米
衣袖展长：98厘米
衣长：119厘米
胸宽：43厘米
腰宽：42厘米
摆宽：55厘米
扣：盘香扣10对，分布在领口3对，
斜襟1对，侧门襟6对；揿纽4对

Coffee half-sleeved silk lined
qipao with woven pattern
Around 1930s-1940s

Collar: Stand-up Collar, 6cm
Length of both sleeves: 98cm
Length: 119cm
Chest width: 43cm
Waist: 42cm
Length of hemline: 55cm
Button: 10 pairs of incense coil frog closures,
3 on collar,
1 on slant opening,
6 on the side;
4 metal press studs

粉色印花棉布无袖单旗袍
约1940年代

领：立领，高3厘米
肩宽：42厘米
衣长：101厘米
胸宽：40厘米
腰宽：37厘米
摆宽：42厘米
侧开衩：11厘米
扣：盘扣3对，分布在领口1对，
斜襟1对，侧门襟1对；揿纽4对

Pink cotton sleeveless
qipao with printed pattern
Around 1940s

Collar: Stand-up Collar, 3cm
Shoulder: 42cm
Length: 101cm
Chest width: 40cm
Waist: 37cm
Length of hemline: 42cm
Side slits: 11cm
Button: 3 pairs of frog closures,
1 on collar, 1 on slant opening, 1 on the side
4 metal press studs

红色印花植绒长袖单旗袍
约1940年代

领：立领，高4厘米
衣袖展长：150厘米
衣长：130厘米
胸宽：45厘米
腰宽：43厘米
摆宽：54厘米
侧开衩：23厘米
扣：盘扣11对，分布在领口2对，
斜襟1对，侧门襟8对

Red flocked qipao with long sleeves and printed pattern
Around 1940s

Collar: Stand-up Collar, 4cm
Length of both sleeves: 150cm
Length: 130cm
Chest width: 45cm
Waist: 43cm
Length of hemline: 54cm
Side slits: 23cm
Button: 11 pairs of frog closures, 2 on collar, 1 on slant opening, 8 on the side

藏青色蕾丝无袖单旗袍
约1940年代

领：立领，高4厘米
肩宽：45厘米
衣长：129厘米
胸宽：52厘米
腰宽：52厘米
摆宽：52厘米
侧开衩：38厘米
扣：盘花扣2对，分布在领口1对，斜襟1对；
盘扣1对，在侧门襟；揿纽7对

Navy blue sleeveless
qipao with lace
Around 1940s

Collar: Stand-up Collar, 4cm
Shoulder: 45cm
Length: 129cm
Chest width: 52cm
Waist: 52cm
Length of hemline: 52cm
Side slits: 38cm
Button: 2 pairs of floral frog closures,
1 on collar, 1 on slant opening;
9 pairs of frog closures on the side;
7 metal press studs

绿地提花真丝无袖单旗袍
约1940年代

领：立领，高4厘米
肩宽：48厘米
衣长：118厘米
胸宽：40厘米
腰宽：39厘米
摆宽：46厘米
侧开衩：28厘米
扣：盘花扣3对，分布在领口1对，斜襟1对，
侧门襟1对；揿纽7对

Green ground sleeveless silk
qipao with woven pattern
Around 1940s

Collar: Stand-up Collar, 4cm
Shoulder: 48cm
Length: 118cm
Chest width: 40cm
Waist: 39cm
Length of hemline: 46cm
Side slits: 28cm
Button: 3 pairs of floral frog closures,
1 on collar,
1 on slant opening;
7 metal press studs

紫地印花丝绒长袖夹旗袍
约1940年代

领：立领，高6.5厘米
衣袖展长：112厘米
衣长：121厘米
胸宽：41厘米
腰宽：40厘米
摆宽：54厘米
侧开衩：28厘米
扣：盘香扣10对，分布在领口3对，
斜襟1对，侧门襟6对

Purple velvet lined qipao with
long sleeves and printed pattern
Around 1940s

Collar: Stand-up Collar, 6.5cm
Length of both sleeves: 112cm
Length: 121cm
Chest width: 41cm
Waist: 40cm
Length of hemline: 54cm
Side slits: 28cm
Button: 10 pairs of incense coil frog closures,
3 on collar,
1 on slant opening,
6 on the side

紫色提花缎面长袖单旗袍
约1930-1940年代

领：立领，高2.5厘米
衣袖展长：108厘米
衣长：111厘米
胸宽：43厘米
腰宽：41厘米
摆宽：45厘米
侧开衩：16厘米
扣：盘花扣3对，分布在领口1对，斜襟1对，
侧门襟1对；揿纽7对

Purple damask long-sleeved
qipao with woven pattern
Around 1930s-1940s

Collar: Stand-up Collar, 2.5cm
Length of both sleeves: 108cm
Length: 111cm
Chest width: 43cm
Waist: 41cm
Length of hemline: 45cm
Side slits: 16cm
Button: 3 pairs of floral frog closures,
1 on collar, 1 on slant opening,
1 on the side;
7 metal press studs

黑色提花真丝长袖单旗袍
约1940年代

领：立领，高6厘米
衣袖展长：129厘米
衣长：119厘米
胸宽：43厘米
腰宽：40厘米
摆宽：44厘米
侧开衩：26厘米
扣：盘花扣3对，分布在领口1对
斜襟1对，侧门襟1对；揿纽6对

Black long-sleeved silk qipao with woven pattern
Around 1940s

Collar: Stand-up Collar, 6cm
Length of both sleeves: 129cm
Length: 119cm
Chest width: 43cm
Waist: 40cm
Length of hemline: 44cm
Side slits: 26cm
Button: 3 pairs of floral frog closures
1 on collar, 1 on slant opening, 1 on the side
6 metal press studs

黑地提花真丝长袖夹旗袍
约1940年代

领：立领，高3厘米
衣袖展长：121厘米
衣长：116厘米
胸宽：42厘米
腰宽：39厘米
摆宽：44厘米
侧开衩：17厘米
扣：盘扣3对，分布在领口1对
斜襟1对，侧门襟1对；揿纽5对

Black silk lined qipao with long sleeves and woven pattern
Around 1940s

Collar: Stand-up Collar, 3cm
Length of both sleeves: 121cm
Length: 116cm
Chest width: 42cm
Waist: 39cm
Length of hemline: 44cm
Side slits: 17cm
Button: 3 pairs of frog closures, 1 on collar,
1 on slant opening, 1 on the side;
5 metal press studs

粉色提花缎面长袖夹旗袍
约1940年代

领：立领，高4.5厘米
衣袖展长：127厘米
衣长：119厘米
胸宽：46厘米
腰宽：42厘米
摆宽：46厘米
侧开衩：22厘米
扣：领口1对钩子，侧门襟盘扣1对，撳纽4对，拉链1条

Pink damask lined qipao with long sleeves and woven pattern
Around 1940s

Collar: Stand-up Collar, 4.5cm
Length of both sleeves: 127cm
Length: 119cm
Chest width: 46cm
Waist: 42cm
Length of hemline: 46cm
Side slits: 22cm
Button: 1 hook-and-eye clasp on the collar, 1 pair of frog closures on slant opening, 4 metal press studs, 1 zip

绿色提花印花真丝无袖单旗袍
约1930-1940年代

领：立领，高3.5厘米
肩宽：47厘米
衣长：133厘米
胸宽：46厘米
腰宽：42厘米
摆宽：48厘米
侧开衩：37厘米
扣：盘花扣3对，分布在领口1对，
斜襟1对，侧门襟1对；揿纽4对

Green silk sleeveless qipao
with woven and printed pattern
Around 1930s-1940s

Collar: Stand-up Collar, 3.5cm
Shoulder: 47cm
Length: 133cm
Chest width: 46cm
Waist: 42cm
Length of hemline: 48cm
Side slits: 37cm
Button: 3 pairs of floral frog closures,
1 on collar, 1 on slant opening,
1 on the side;
4 metal press studs

满地印花真丝短袖夹旗袍
约1930-1940年代

领：立领，高3.5厘米
肩宽：47厘米
衣长：127厘米
胸宽：46厘米
腰宽：37厘米
摆宽：45厘米
侧开衩：40厘米
扣：盘扣3对，分布在领口1对，
斜襟1对，侧门襟1对；揿纽6对

Silk short-sleeved lined
qipao with full printed pattern
Around 1930s-1940s

Collar: Stand-up Collar, 3.5cm
Shoulder: 47cm
Length: 127cm
Chest width: 46cm
Waist: 37cm
Length of hemline: 45cm
Side slits: 40cm
Button: 3 pairs of frog closures,
1 on collar, 1 on slant opening,
1 on the side;
6 metal press studs

提花印花真丝无袖单旗袍
约1940年代

领：立领，高3.5厘米
肩宽：46厘米
衣长：106厘米
胸宽：41厘米
腰宽：39厘米
摆宽：47厘米
侧开衩：9厘米
扣：盘扣1对，在侧门襟；揿纽4对

Silk sleeveless qipao with
woven and printed pattern
Around 1940s

Collar: Stand-up Collar, 3.5cm
Shoulder: 46cm
Length: 106cm
Chest width: 41cm
Waist: 39cm
Length of hemline: 47cm
Side slits: 9cm
Button: 1 pair of frog closures on slant opening;
4 metal press studs

印花中袖夹旗袍
约1930-1940年代

领：立领，高4.5厘米
衣袖展长：90厘米
衣长：116厘米
胸宽：37厘米
腰宽：36厘米
摆宽：50厘米
侧开衩：33厘米
扣：盘扣12对，分布在领口2对，斜襟1对，侧门襟9对

Half-sleeved lined qipao with printed pattern
Around 1930s-1940s

Collar: Stand-up Collar, 4.5cm
Length of both sleeves: 90cm
Length: 116cm
Chest width: 37cm
Waist: 36cm
Length of hemline: 50cm
Side slits: 33cm
Button: 12 pairs of frog closures, 2 on collar, 1 on slant opening, 9 on the side

满地提花真丝长袖夹旗袍
约1940年代

领：立领，高4厘米
衣袖展长：140厘米
衣长：120厘米
胸宽：48厘米
腰宽：42厘米
摆宽：48厘米
侧开衩：17厘米
扣：盘扣3对，分布在领口1对，斜襟1对，侧门襟1对；揿纽5对

Silk long-sleeved lined qipao with full woven pattern
Around 1940s

Collar: Stand-up Collar, 4cm
Length of both sleeves: 140cm
Length: 120cm
Chest width: 48cm
Waist: 42cm
Length of hemline: 48cm
Side slits: 17cm
Button: 3 pairs of frog closures, 1 on collar, 1 on slant opening, 1 on the side; 5 metal press studs

蓝色提花真丝长袖夹旗袍
约1940年代

领：立领，高5.5厘米
衣袖展长：130厘米
衣长：122厘米
胸宽：45厘米
腰宽：44厘米
摆宽：60厘米
侧开衩：43厘米
扣：盘香扣10对，分布在领口3对，斜襟1对，侧门襟6对

Blue silk long-sleeved lined qipao with woven pattern
Around 1940s

Collar: Stand-up Collar, 5.5cm
Length of both sleeves: 130cm
Length: 122cm
Chest width: 45cm
Waist: 44cm
Length of hemline: 60cm
Side slits: 43cm
Button: 10 pairs of incense coil frog closures, 3 on collar, 1 on slant opening, 6 on the side

蓝地提花真丝长袖夹旗袍
约1940年代

领：立领，高3.5厘米
衣袖展长：133厘米
衣长：122厘米
胸宽：44厘米
腰宽：39厘米
摆宽：45厘米
侧开衩：18厘米
扣：盘扣3对，分布在领口1对，斜襟1对，侧门襟1对；揿纽4对

Blue silk lined qipao with long sleeves and woven pattern
Around 1940s

Collar: Stand-up Collar, 3.5cm
Length of both sleeves: 133cm
Length: 122cm
Chest width: 44cm
Waist: 39cm
Length of hemline: 45cm
Side slits: 18cm
Button: 3 pairs of frog closures, 1 on collar, 1 on slant opening, 1 on the side;
4 metal press studs

图书在版编目（CIP）数据

衷藏雅尚 · 海上流晖：王水衷捐赠服饰精选／上海市历史博物馆，上海革命历史博物馆编．－－上海：上海书画出版社，2024.9．－－ISBN 978－7－5479－3437－1

Ⅰ．TS941.717－64

中国国家版本馆CIP数据核字第2024VH1950号

衷藏雅尚·海上流晖:王水衷捐赠服饰精选

上海市历史博物馆／上海革命历史博物馆 编

责任编辑 孙 晖 凌云之君 赖 妮 袁 媛
审 读 田松青
封面设计 朱晟昊
技术编辑 顾 杰

出版发行 上海世纪出版集团 上海书画出版社
地 址 上海市闵行区号景路159弄A座4楼 201101
网 址 www.shshuhua.com
E－mail shuhua@shshuhua.com
设 计 上海金脉美术设计有限公司
印 刷 上海艾登印刷有限公司
经 销 各地新华书店
开 本 889×1194 1/16
印 张 11
版 次 2024年9月第1版 2024年9月第1次印刷

书 号 ISBN 978－7－5479－3437－1
定 价 380.00元

若有印刷、装订质量问题，请与承印厂联系